SpringerBriefs in Applied Sciences and Technology

Computational Intelligence

Series Editor

Janusz Kacprzyk, Systems Research Institute, Polish Academy of Sciences, Warsaw, Poland

SpringerBriefs in Computational Intelligence are a series of slim high-quality publications encompassing the entire spectrum of Computational Intelligence. Featuring compact volumes of 50 to 125 pages (approximately 20,000–45,000 words), Briefs are shorter than a conventional book but longer than a journal article. Thus Briefs serve as timely, concise tools for students, researchers, and professionals.

More information about this series at http://www.springer.com/series/10618

Patricia Melin · Gabriela E. Martinez

Extension of the Fuzzy Sugeno Integral Based on Generalized Type-2 Fuzzy Logic

 Springer

Patricia Melin
Division of Graduate Studies
and Research
Tijuana Institute of Technology
Tijuana, Baja California, Mexico

Gabriela E. Martinez
Division of Graduate Studies
and Research
Tijuana Institute of Technology
Tijuana, Baja California, Mexico

ISSN 2191-530X   ISSN 2191-5318   (electronic)
SpringerBriefs in Applied Sciences and Technology
ISSN 2625-3704   ISSN 2625-3712   (electronic)
SpringerBriefs in Computational Intelligence
ISBN 978-3-030-16415-7   ISBN 978-3-030-16416-4   (eBook)
https://doi.org/10.1007/978-3-030-16416-4

Library of Congress Control Number: 2019935488

This Springer imprint is published by the registered company Springer Nature Switzerland AG
The registered company address is: Gewerbestrasse 11, 6330 Cham, Switzerland

Preface

In this book, an extension of the aggregation operator of the generalized interval type-2 Sugeno integral by means of generalized type-2 fuzzy logic is presented. The main goal of extending the Sugeno integral aggregation operator, is to give it the ability to handle higher levels of uncertainty by adding any number of sources and types of information in a wide variety of applications. In addition, for demonstrating that the extended aggregation operator with generalized type-2 fuzzy logic presents a better performance than the traditional operator or extended operator with interval type-2 fuzzy logic. As part of the objectives, the proposed method was implemented in a modular neural network applied to face recognition and in an edge detector. Comparisons of the results obtained with respect to other aggregation operators were also made. During the development of the extension of the operators of a generalized type-2 fuzzy system were applied to the fuzzy densities, the lambda calculation, fuzzy measurements and the fuzzy Sugeno integral.

This method offers the best performance in applications of pattern recognition and edge detection, but the extension of the Sugeno integral can be used in any application where it is necessary to add numerical information.

This book can be considered as a reference for future research in which are involved operators that use some type of measure and is organized as follows: In Chap. 1, we begin by offering a brief introduction to the potential use of the aggregation operators in real applications.

We describe in Chap. 2, the basic concepts of type-1 fuzzy logic, interval type-2 and generalized type-2 fuzzy logic are presented. Also we explained the Sugeno integral aggregation operator and the Interval type-2 Sugeno integral, which serve as a reference for developing the proposed method. Additionally, we explain in detail the concepts of Modular neural networks and edge detectors. The methodology used for developing the generalized type-2 Sugeno integral, is presented in Chap. 3. In Chap. 4 offers presents simulation results with benchmark faces databases and benchmark images to illustrate the advantages of the proposed generalized type-2 Sugeno integral method and finally, Chap. 5 presents the conclusions and future work from the obtained results.

Tijuana, Mexico

Patricia Melin
Gabriela E. Martinez

Contents

List of Figures

List of Tables

Keywords

Sugeno integral · Fuzzy measures · Fuzzy logic ·
Generalized interval type-2 fuzzy logic · Modular neural network ·
Face recognition · Edge detection

Chapter 1
Introduction to the Type-2 Fuzzy Sugeno Integral

The process of information aggregation is a key element in any system in which it is necessary to perform a decision making task. Frequently, the result aggregated information considerably reduces the quantity of original information; however, performing it efficiently is one of the main tasks of various systems that handling large amounts of information, whose quality and accuracy can be quite different.

In previous work we have consider operators that do not adequately reflect the process of aggregation of different sources or different criteria, so it is more convenient to use more robust aggregators, which can be able to handle higher degrees of uncertainty, and some of them are also capable of handling weights. The main goal of aggregation operators is for combining information when they can be mathematically formalized. Aggregation operators are aimed at reducing a set of numbers into a unique representative value, as we can notice in Fig. 1.1. It is important to note that any aggregation or fusion process is based on numerical aggregation. An operator considers that the input variables are the information sources to combine and the output is the aggregation of the results.

In the current bibliography we can find a variety of aggregation operators that have been proposed, and the most traditional ones are: the geometric mean, arithmetic mean, weighted arithmetic mean, harmonic mean. However, there are other more complex operators that have been put forward that use measures or weights, for instance the ordered weighted averaging (OWA) [1, 2], weighted OWA (WOWA) [3], Choquet Integral [4] and Sugeno Integral [5, 6].

The main contribution of this work is the proposed extension of the Sugeno integral based on the operators of generalized type-2 fuzzy logic, by means of which one can achieve a better handling of the uncertainty than with the existing aggregation operators (only using type-1 fuzzy logic or crisp values).

The emergence of type-1 fuzzy systems [7–9] has allowed a great advance in various engineering applications, which has increased even more with the appearance of interval type-2 fuzzy systems [10–12] and lately with generalized type-2 fuzzy systems [13–17]. The idea of using interval type-2 fuzzy systems and generalized type-2 fuzzy systems is that they can have a greater ability for handling of uncertainty,

© The Author(s), under exclusive license to Springer Nature Switzerland AG 2020
P. Melin and G. E. Martinez, *Extension of the Fuzzy Sugeno Integral Based on Generalized Type-2 Fuzzy Logic*, SpringerBriefs in Computational Intelligence,
https://doi.org/10.1007/978-3-030-16416-4_1

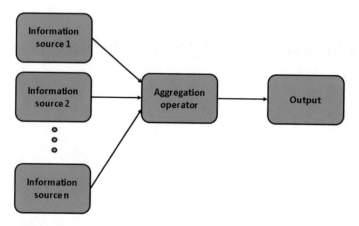

Fig. 1.1 Representation of the aggregation operator

so it is expected that better results will be obtained than with traditional type-1 fuzzy systems. However, generalized type-2 fuzzy systems require a lot of computing power to carry out all their calculations. There are some methods that have been proposed that enable reducing the computing overhead of generalized systems, such as the α-planes [14–16] and zSlices [15, 17]. Recently, many applications of type-2 fuzzy logic [18–21] have been put forward and could be improved with the proposed method in this Book. The main contribution of this research work is to extend the Sugeno integral using generalized and interval type-2 fuzzy logic systems, which enables better handling of uncertainty in decision making.

References

1. Yager RR (1988) On ordered weighted averaging aggregation operators in multi-criteria decision making. IEEE Trans Syst Man Cybern 18:183–190
2. Yager RR, Kacprzyk J (1997) The ordered weighted averaging operators, theory and applications. Springer Science Business Media, LLC
3. Larsen HL (1999) Importance weighted OWA aggregation of multicriteria queries. Fuzzy Inf Process Soc
4. Choquet G (1953) Theory of capacities. Ann Inst Fourier 5:131–295
5. Sugeno M (1974) Theory of fuzzy integrals and its applications, Doctoral Thesis, Tokyo Institute of Technology
6. Melin P, Mancilla A, Lopez M, Mendoza O (2007) A hybrid modular neural network architecture with fuzzy Sugeno integration for time series forecasting. Appl Soft Comput 7(4):1217–1226
7. Zadeh LA (1988) Fuzzy logic. Computer 1(4):83–93
8. Zadeh LA (1975) The concept of a linguistic variable and its application to approximate reasoning. Inf Sci 8(3):199–249
9. Zadeh LA (1965) Fuzzy sets. Inf Control 8:338–353
10. Karnik NN, Mendel JM, Liang Q (1999) Type-2 fuzzy logic systems. IEEE Trans Fuzzy Syst 7(6):643–658

11. Liang Q, Mendel J (2000) Interval type-2 fuzzy logic systems: theory and design. IEEE Trans Fuzzy Syst 8(5):535–550
12. Mendel J (2001) Uncertain rule-based fuzzy logic systems: introduction and new directions. Prentice-Hall
13. Liu F (2008) An efficient centroid type-reduction strategy for general type-2 fuzzy logic system. Inf Sci 178(9):2224–2236
14. Mendel JM, Fellow L, Liu F, Zhai D (2009) α-plane representation for type-2 fuzzy sets. theory and applications. IEEE Trans Fuzzy Syst 17(5):1189–1207
15. Wagner C, Hagras H (2010) Toward general type-2 fuzzy logic systems based on zSlices. Fuzzy Syst IEEE Trans 18(4):637–660
16. Mendel JM (2010) Comments on alpha-plane representation for type-2 fuzzy sets: theory and applications. IEEE Trans Fuzzy Syst 18(1):229–230
17. Wagner C, Hagras H (2011) Employing zSlices based general type-2 fuzzy sets to model multilevel agreement. In: 2011 IEEE symposium on advances in type-2 fuzzy logic systems, pp 50–57
18. Melin P, González CI, Castro JR, Mendoza O, Castillo O (2014) Edge-detection method for image processing based on generalized type-2 fuzzy logic. IEEE Trans Fuzzy Syst 22(6):1515–1525
19. González CI, Melin P, Castro JR, Castillo O, Mendoza O (2016) Optimization of interval type-2 fuzzy systems for image edge detection. Appl Soft Comput 47:631–643
20. González CI, Melin P, Castro JR, Mendoza O, Castillo O (2016) An improved sobel edge detection method based on generalized type-2 fuzzy logic. Soft Comput 20(2):773–784
21. Ontiveros E, Melin P, Castillo O (2018) High order α-planes integration: a new approach to computational cost reduction of General Type-2 Fuzzy Systems. Eng Appl AI 74:186–197

Chapter 2
Basic Theory for the Type-2 Fuzzy Sugeno Integral

This work is based on the aggregation operator of the Sugeno integral which makes use of Sugeno measures, and with this enables it to have the ability of uncertainty management. However, the main contribution consists in extending the operator through the use of generalized type-2 fuzzy logic and α planes to include managing information with uncertainty. In this chapter we present some basic concepts needed to better understand the general idea and the context of the work presented in this Book.

2.1 Fuzzy Logic

In a great variety of real applications, when it is necessary to manipulate information, is common to find problems due to the presence of uncertainty, which occurs when using inaccurate or imprecise data. Zadeh in 1965 proposed the solution to this problem by giving the definition of a fuzzy set [1], and to complete the solution, Michio Sugeno introduced the terms of fuzzy measure and fuzzy integral [2] as the most appropriate way to measure a certain degree of uncertainty, and these values depend only on human subjectivity. Fuzzy logic was originally defined by Zadeh in 1965.

2.1.1 Type-1 Fuzzy Sets

If X is a collection of objects denoted by x, then a "Fuzzy set" A in X, can be represented by Eq. (2.1) and is defined as a set of ordered pairs

© The Author(s), under exclusive license to Springer Nature Switzerland AG 2020
P. Melin and G. E. Martinez, *Extension of the Fuzzy Sugeno Integral Based
on Generalized Type-2 Fuzzy Logic*, SpringerBriefs in Computational Intelligence,
https://doi.org/10.1007/978-3-030-16416-4_2

$$A = \{(x, \mu_A(x)) | \forall x \in X\} \tag{2.1}$$

where $\mu_A(x)$ represent the membership function (MF) of the fuzzy set A.

2.1.2 Type-2 Fuzzy Sets

A Type-2 fuzzy set, denoted by $\tilde{A}$, is characterized by a Type-2 membership function $\mu_{\tilde{A}}(x, u)$, where $x \in X$ and $u \in J_x \subseteq [0, 1]$, and can be expressed by (2.2) in which $0 \leq \mu_{\tilde{A}}(x, u) \leq 1$.

$$\tilde{A} = \left\{((x, u), \mu_{\tilde{A}}(x, u)) | \forall x \in X, \forall u \in J_x \subseteq [0, 1]\right\} \tag{2.2}$$

$\tilde{A}$ can also be expressed in an equivalent expression,

$$\tilde{A} = \int_{x \in X} \int_{x \in j_x} \frac{\mu_{\tilde{A}}(x, u)}{x, u} J_x \subseteq [0, 1] \tag{2.3}$$

it can be denoted by Eq. (2.3), where $\int \int$ denotes the union over all admissible x and u [18].

The footprint of uncertainty (FOU) can be described as the region limited by upper and lower membership functions. An upper membership function and a lower membership function [3] are two Type-1 membership functions that bound the FOU of an interval Type-2 fuzzy set $\tilde{A}$. The upper membership function is associated with the upper bound of the FOU $(\tilde{A})$ is represented by Eq. (2.4), and is denoted by $\bar{\mu}_{\tilde{A}}(x), \forall x \in X$.

$$\bar{\mu}_{\tilde{A}}(x) \equiv \overline{FOU(\tilde{A})} \quad \forall x \in X \tag{2.4}$$

A lower membership function is associated with lower bound of the FOU $(\tilde{A})$, it can be expressed by (2.5) and is denoted by $\underline{\mu}_{\tilde{A}}(x), \forall x \in X$

$$\underline{\mu}_{\tilde{A}}(x) \equiv \underline{FOU(\tilde{A})} \quad \forall x \in X \tag{2.5}$$

because the domain of a secondary membership function has been constrained in [0, 1], the lower and upper membership functions always exist [3].

$$FOU(\tilde{A}) = \bigcup_{\forall x \in X} \left[\underline{\mu}_{\tilde{A}}(x), \bar{\mu}_{\tilde{A}}(x)\right] \tag{2.6}$$

The FOU ($\tilde{A}$) of an interval type-2 membership function can also be expressed by Eq. (2.6).

2.1.3 Generalized Type-2 Fuzzy Logic

Definition: a Generalized type-2 fuzzy set denoted by $\tilde{A}$, is characterized by a type-2 membership function $\mu_{\tilde{A}}(x, u)$, where $x \in X$, $u \in J_x^u \subseteq [0, 1]$ and $0 \leq \mu_{\tilde{A}}(x, u) \leq 1$ and can be represented by (2.7) [4–8].

$$\tilde{A} = \{(x, \mu_{\tilde{A}}(x)) | x \in X\} = \{((x, u), \mu_{\tilde{A}}(x, u)) | \forall x \in X, \forall u \in J_x^u \subseteq [0, 1]\} \quad (2.7)$$

if $\tilde{A}$ is continuous is denoted by Eq. (2.8).

$$
\begin{aligned}
\tilde{A} &= \left\{ \int_{x \in X} \frac{\mu_{\tilde{A}}(x)}{x} \right\} \\
&= \left\{ \int_{x \in X} \int_{u \in J_x^u \subseteq [0,1]} \mu_{\tilde{A}}(x, u)/(x, u) \right\} \\
&= \left\{ \int_{x \in X} \left[\int_{u \in J_x^u \subseteq [0,1]} f_x(u)/u \right] /x \right\} \quad (2.8)
\end{aligned}
$$

where $\iint$ represents the union for x and u. x is primary domain, J_x is the secondary domain, $\mu_{\tilde{A}}(x)$ is the secondary membership function at x is called a vertical slice of $\tilde{A}$ [6] and all secondary grades are represented by $\mu_{\tilde{A}}(x, u) \in [0, 1]$; and if $\tilde{A}$ is discrete is denoted by Eq. (2.9).

$$
\begin{aligned}
\tilde{A} &= \left\{ \sum_{x \in X} \mu_{\tilde{A}}(x)/x \right\} \\
&= \left\{ \sum_{x \in X} \sum_{u \in J_x^u \subseteq [0,1]} \mu_{\tilde{A}}(x, u)/(x, u) \right\} \\
&= \left\{ \sum_{x \in X} \left[\sum_{u \in J_x^u \subseteq [0,1]} f_x(u)/u \right] /x \right\} \\
&= \left\{ \sum_{i=1}^{N} \left[\sum_{k=1}^{M_i} f_{x_i}(u_{ik})/u_{ik} \right] /x_i \right\} \quad (2.9)
\end{aligned}
$$

where $\Sigma\Sigma$ denotes the union of x and u.

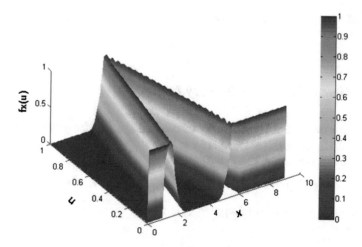

Fig. 2.1 Generalized type-2 membership function

Fig. 2.2 FOU of a general
type-2 membership function

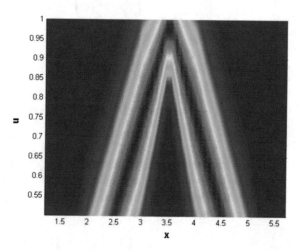

We can visualize the representation of a generalized type-2 membership function
in Fig. 2.1, which is associated with the third dimension; the footprint of uncertainty
(FOU) of the function can be found in Fig. 2.2.

The generalized type-2 fuzzy systems have been successfully applied in [9–12].

2.1.3.1   Alpha Planes

An α plane for a generalized type-2 fuzzy set $\tilde{A}$ is denoted by $\tilde{A}_{\alpha}$ and can be defined
as the union of all primary membership functions of $\tilde{A}$, where the secondary mem-

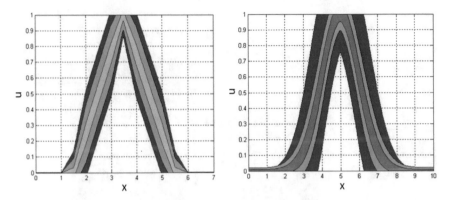

Fig. 2.3 α plane in generalized type-2 membership functions

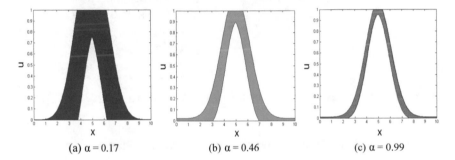

(a) α = 0.17   (b) α = 0.46   (c) α = 0.99

Fig. 2.4 Cuts at certain points of the GMF

bership degrees are greater or equal to α ($0 \leq \alpha \leq 1$) and can be represented by (2.10) [4, 5], and can be graphically illustrated in Fig. 2.4.

$$\tilde{A}_\alpha = \left\{ (x, u), \mu_{\tilde{A}}(x, u) \geq \alpha | \forall x \in X, \forall u \in J_x \subseteq [0, 1] \right\}$$

$$= \int_{\forall x \in X} \int_{\forall u \in J_x} \{(x, u) | fx(u) \geq \alpha\} \tag{2.10}$$

where the union of all the α alpha planes is represented by (2.11), and $R_{\tilde{A}_\alpha}$ represents a horizontal slice. In Fig. 2.3 we can appreciate the α alpha planes made for generalized membership functions.

Also in Fig. 2.4 we can appreciate the representation of the three α planes in different points, the first α plane was realized in the point 0.17 (blue), the second α plane was in 0.46 (green) and the third in the point 0.99 (red).

$$\tilde{A} = \bigcup_{\alpha \in [0, 1]} R_{\tilde{A}_\alpha} \tag{2.11}$$

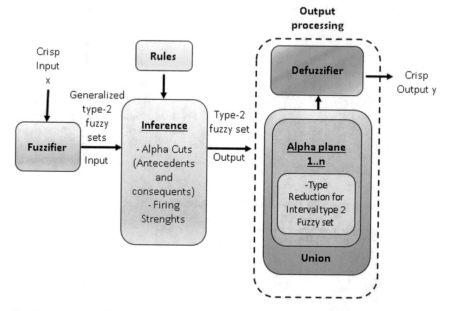

Fig. 2.5 Diagram of Generalized type-2 fuzzy logic system based on α planes

2.1.3.2 Generalized Type-2 Fuzzy Systems Based on α-Planes

Due to the computational complexity of a generalized type-2 fuzzy system, the calculation is carried out by means of an approximation using α planes or zSlices. The main idea of this work is based on the operators of a generalized type-2 fuzzy system based on α planes.

In Fig. 2.5 is presented a diagram where we can visualize the generalized type-2 fuzzy system, and here we can appreciate its main elements, which are: the fuzzifier process, fuzzy rules, inference, type reducer, and the process of defuzzification; this representation is based on the approximation of α planes.

2.1.3.3 Parametrization of a Generalized Type-2 Membership Function (GT2MF)

In this section we define a generalized triangular membership function in the primary with a Gaussian function in the secondary. In Fig. 2.6 (a) we can appreciate the interval type-2 membership function with the respective parameters and in Fig. 2.6 (b) can be observed the membership function represented in the third dimension. The parameters of the membership functions were obtained by (2.11) and we defined the generalized membership function as follows:

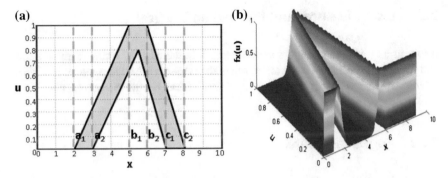

Fig. 2.6 Interval type-2 and generalized type-2 membership functions

$$\bar{\mu}(x, u) = trigausstype2(x, u, [a_1, b_1, c_1, a_2, b_2, c_2, \rho])$$

where x is the partition of the primary membership function and u represents the domain of the secondary membership function, and is calculated with (2.12)

$$\tilde{\mu}(x, u) = \exp\left[-\frac{1}{2}\left(\frac{u - px}{\sigma_u}\right)^2\right] \qquad (2.12)$$

in which σ_u is calculated using (2.13)

$$\sigma_u = \frac{1 + \rho}{2\sqrt{3}}\delta + \varepsilon \qquad (2.13)$$

where ρ, is the fraction of uncertainty of the support secondary membership function and using (2.14) δ is calculated as

$$\delta = \bar{\mu}(x) - \underline{\mu}(x) \qquad (2.14)$$

with (2.15) px is determined by

$$px = \max\left(\min\left(\frac{x - a}{b - a}, \frac{c - x}{c - b}\right), 0\right) \qquad (2.15)$$

and the parameters a, b and c are calculated with the use of (2.16)

$$a = \frac{a_1 + a_2}{2}, \quad b = \frac{b_1 + b_2}{2}, \quad c = \frac{c_1 + c_2}{2} \qquad (2.16)$$

where "trigausstype2" stands for the Gaussian interval type-2 membership function with uncertain mean. This generalized membership function was the one used for the development of the proposed method.

2.2   Fuzzy Measures and the Sugeno Integral

As far as aggregation operators are concerned, in the last few years, fuzzy integrals and fuzzy measures have had a big boom in the research area that is why it is of great interest to work with this type of operators which can handle measurements.

2.2.1   Sugeno Measures

A special type of monotonic measures are the Sugeno λ-measures [13, 14] defined as follows: If we have a finite set $x = \{x_1, x_2, ..., x_n\}$, a fuzzy measure μ with respect to the dataset X, can be defined as a function $\mu : 2^x \to [0, 1]$ that must fulfill the following conditions:

(1) $\mu(X) = 1; \mu(\emptyset) = 0$ (Boundary conditions)
(2)   If $A \subseteq B$, then $\mu(A) \le \mu(B)$ (Monotonicity)

where A and B are subsets of X.

A Sugeno measure is a fuzzy measure or λ-fuzzy, if it satisfies the condition (1) of addition for some $\lambda > -1$.

$$\mu(A \cup B) = \mu(A) + \mu(B) + \lambda\, \mu(A)\mu(B) \qquad (2.17)$$

Equation (2.17) is usually called the λ-rule. When X is a finite set and the values $\mu(\{x\})$, called fuzzy densities, are given for each $x \in X$, these densities are interpreted as the importance or relevance of the individual information sources. The measure of a set A of the information sources is interpreted as the importance of that subset of sources towards answering or solving a particular question or problem [15].

The value of $\mu(A)$ for each $A \subset P(X)$, can be determined by the recurrent application of the λ-rule. This value can be expressed in the following way.

$$\mu(A) = \left[\prod_{x \in A} (1 + \lambda\mu(\{x\})) \right] / \lambda \qquad (2.18)$$

We can notice that once the values of the fuzzy densities $\mu(\{x\})$ are assigned for each $x \in X$, the value of λ can be calculated by using the constraint $\mu(\{x\}) = 1$. So that by applying this restriction (2.18) we obtain (2.19)

$$\lambda + 1 = \prod_{i=1}^{n} (1 + \lambda\mu(\{x_i\})) \qquad (2.19)$$

Once the densities are known, the λ parameter can be computed using (2.19) and is specific to this class of measures.

Sugeno proved that this polynomial has a real root greater than -1 and several researchers have observed that this polynomial equation is easily solved numerically. By property (2.17), specifying the n different densities, thereby reducing the number of free parameters from $2^n - 2$ to n [13], the value of parameter λ is determined with the help of the following theorem [16]:

Theorem 2.1 Let $\mu(\{x\}) < 1$ for each $x \in X$ and let $\mu(\{x\}) > 0$ for at least two elements of X. Then (2.19) determines a unique parameter λ in the following way.

- If $\sum_{x \in X} \mu(\{x\}) < 1$, then λ is equal to a unique root of the equation in the interval $(0, \infty)$.
- If $\sum_{x \in X} \mu(\{x\}) = 1$, then $\lambda = 0$; that is the unique root of the equation.
- If $\sum_{x \in X} \mu(\{x\}) > 1$, then λ is equal to a unique root of the equation in the interval $(-1, 0)$.

When μ is a λ-fuzzy measure, the values of $\mu(A_i)$ can be computed by means of (2.18), or recursively, after a descendent reordering of the sets X and $\mu(\{x\})$, with respect to the values of the elements of set X [17].

There are two types of Integral that can perform the calculation of Sugeno measures, these are the Sugeno Integral (SI) and the Choquet Integral, which have already been used in a variety of applications [18].

2.2.2 Sugeno Integral

Using the concept of fuzzy measures, Sugeno also proposed the concept of fuzzy Integrals as nonlinear functions defined with respect to fuzzy measures such as the λ-fuzzy measure. The Sugeno Integral (SI) generalizes the "max-min" operator. One can interpret the fuzzy Integral as finding the maximum degree of similarity between the target and the expected value as shown in (2.20).

$$Sugeno_\mu(x_1, x_2, \ldots, x_n) = \max_{i=1\ldots n} \left(\min \left(D(x_{\sigma(i)}), \mu(A_{\sigma(i)}) \right) \right) \qquad (2.20)$$

where $x_{\sigma(i)}$ indicates that the indices of the data $D(x_{\sigma(i)})$ have been permuted as $0 \leq D(x_{\sigma(1)}) \leq D(x_{\sigma(2)}) \leq \ldots \leq D(x_{\sigma(n)}) \leq 1$, and where $A_{\sigma(i)} = \{A_{\sigma(1)}, \ldots, A_{\sigma(n)}\}$.

The Sugeno Integral can be used to solve problems that consider a finite set of n elements $X = \{x_1, \ldots, x_n\}$.

2.2.3   Fuzzy Measures and Sugeno Integral for Interval Type-2 Fuzzy Sets

In this section the interval type-2 Sugeno integral (IT2SI) is defined.

2.2.3.1   Fuzzy Measures with Interval Type-2 Fuzzy Sets

To calculate the fuzzy density of each information source, it takes the maximum value of each X_i and then a footprint of uncertainty (FOU) is added to form a Type-2 fuzzy set. We need to add a FOU to create a density based on a fuzzy interval. Equation (2.21) can be used to approximate the center of the interval and Eqs. (2.22) and (2.23) for calculate the values of the upper and lower interval for each fuzzy density. Note that the domains for $\mu_U(x_i)$ and $\mu_L(x_i)$ are given in Theorem 2.1 [19].

The calculation of the left and right fuzzy densities can be defined as follows:

$$\mu_c(x_i) = \max(X_i) \tag{2.21}$$

$$\mu_U(x_i) = \begin{cases} \mu_c(x_i) - FOU_\mu/2; & \text{if } \mu_c(x_i) > FOU_\mu/2 \\ \varepsilon_U & \text{otherwise} \end{cases} \tag{2.22}$$

$$\mu_L(x_i) = \begin{cases} \mu_c(x_i) + FOU_\mu/2; & \text{if } \mu_c(x_i) < (1 - FOU_\mu/2) \\ \varepsilon_L & \text{otherwise} \end{cases} \tag{2.23}$$

where $i = 2, 3, \ldots, n$, and n represents the number of information sources, μ_c is the central fuzzy density, μ_U is the upper fuzzy density, μ_L is the lower fuzzy density, and FOUμ is the footprint of uncertainty added to the fuzzy densities. The ε_U and ε_L values are determined by the following conditions, in order to satisfy Theorem 2.1:

- ε_U is the smallest number in $(0, \mu_c(x_i))$ depending on the application.
- ε_L is the biggest number in $(\mu_c(x_i), 1)$ depending on the application.

The λ_U upper and λ_L lower parameters for each side of the interval can be calculated with (2.24) and (2.25)

$$\lambda_U + 1 = \prod_{i=1}^{n} (1 + \lambda_U \mu_U(\{x_i\})) \tag{2.24}$$

$$\lambda_L + 1 = \prod_{i=1}^{n} (1 + \lambda_L \mu_L(\{x_i\})) \tag{2.25}$$

Once the λ_U and λ_L are obtained the upper μ_U (A_i) (2.26) (2.27) and the lower $\mu_L(A_i)$ (2.28) (2.29) fuzzy measures can be calculated by extending the recursive formula as follows:

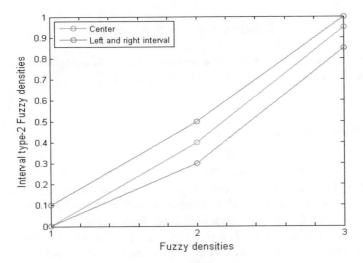

Fig. 2.7 Interval type-2 fuzzy densities for specific values $\mu_{c(x_1)} = 0.02$, $\mu_{c(x_2)} = 0.4$, $\mu_{c(x_3)} = 0.95$ and FOU $\mu = 0.2$

$$\mu_U(A_1) = \mu_U(x_1) \tag{2.26}$$

$$\mu_U(A_i) = \mu_U(x_i) + \mu_U(A_{i-1}) + \lambda_U \mu_U(x_i) \mu_U(A_{i-1}) \tag{2.27}$$

$$\mu_L(A_1) = \mu_L(x_1) \tag{2.28}$$

$$\mu_L(A_i) = \mu_L(x_i) + \mu_L(A_{i-1}) + \lambda_L \mu_L(x_i) \mu_L(A_{i-1}) \tag{2.29}$$

In Fig. 2.7 it can be appreciated the generated interval of the fuzzy densities.

2.2.3.2   Sugeno Integral with Interval Type-2 Fuzzy Sets

If we extend Eq. (2.20) using intervals we obtain the following expression:

$$Sugeno_U = \max_{i=1..n} (\min(D_U(x_i), \mu_U(A_i))) \tag{2.30}$$

$$Sugeno_L = \max_{i=1..n} (\min(D_L(x_i), \mu_L(A_i))) \tag{2.31}$$

where $Sugeno_U$ (2.30) and $Sugeno_L$ (2.31) represent the upper and lower interval extremes of the Sugeno integral.

$$Sugeno = (Sugeno_U + Sugeno_L)/2 \qquad (2.32)$$

Once the upper and lower intervals are calculated, the next step is to calculate the average of both values to obtain the IT2SI using (2.32).

2.3   Edge Detection

One of the main steps that are performed in many applications of digital images processing is the extraction of patterns or significant elements. One of the main features that is usually obtained is the edge or outline of an object of the image, which provides information of great importance for later stages.

Edge detection is a process in digital image analysis that detects changes in light intensity, and it is an essential part of many computer vision systems. The edge detection process is useful for simplifying the analysis of images by dramatically reducing the amount of data to be processed [19]. An edge may be the result of changes in light absorption (shade/color/texture, etc.) and can delineate the boundary between two different regions in an image.

The resultant images of edge detectors preserve more details of the original images, which is a desirable feature for a pattern recognition system. In the literature there are various edge detectors, amongst them, the best known operators are the Morphological gradient, Sobel [20], Prewitt [21], Robert [22], Canny [19] and Kirsch [23]. There are also edge detection methods that combine traditional detection methods with different techniques or even intelligent methods, which have been successful such as type-1 fuzzy systems [24], interval type-2 fuzzy systems combined with the Sobel operator [25], interval type-2 fuzzy systems and the morphological gradient [26, 27], Sugeno integral and interval type-2 Sugeno integral [27], among others. In Table 2.1, a summary of existing edge detection methods is presented.

Table 2.1 Edge detection methods

Traditional methods	Computational intelligence techniques
Morphological gradient	Morphological gradient with fuzzy system
Sobel	Morphological gradient with interval type-2 fuzzy system
Prewitt	
Roberts	Sobel with fuzzy system
Laplacian	
Canny	Sobel with interval type-2 fuzzy system
Kirsch	
LoG	Sugeno integral
Zero crossing	Interval type-2 Sugeno Integral

The edges of a digital image can be defined as transitions between two regions of significantly different gray levels, and they provide valuable information about the boundaries of objects and can be used to recognize patterns and objects, segmenting an image, etc.

2.3.1 Edge Detector Based on Gradient

Gradient operators are based on the idea of using the first or second derivative of the gray level. Based on an image $f(x, y)$, the gradient of point (x, y) is defined as a gradient vector (∇f) by Eq. (2.33) and is calculated as follows:

$$\nabla f = \begin{bmatrix} Gx \\ Gy \end{bmatrix}$$

$$\nabla f = \begin{bmatrix} \frac{\partial f}{\partial x} \\ \frac{\partial f}{\partial y} \end{bmatrix} \tag{2.33}$$

where, the gradient magnitude vector $(mag(\nabla f))$ is calculated with:

$$mag(\nabla f) = \left[\left(\frac{\partial f}{\partial x} \right)^2 + \left(\frac{\partial f}{\partial y} \right)^2 \right]^{1/2}$$

$$mag(\nabla f) = \left[G_x^2 + G_y^2 \right]^{1/2} \tag{2.34}$$

In this case, we are going to use G_i instead of $mag(\nabla f)$, in other words we apply (2.34) for a matrix of 3×3 as it is shown in Fig. 2.8. The values for z_i are obtained using (2.35), and the possible direction of edge G_i with (2.36–2.39). The G gradient can be calculated using (2.40).

$$
\begin{aligned}
z_1 &= (x - 1, y - 1) & z_6 &= (x, y + 1) \\
z_2 &= (x - 1, y) & z_7 &= (x + 1, y - 1) \\
z_3 &= (x - 1, y + 1) & z_8 &= (x + 1, y) \\
z_4 &= (x, y - 1) & z_9 &= (x + 1, y + 1) \\
z_5 &= (x, y)
\end{aligned}
\tag{2.35}
$$

$$G1 = \sqrt{(z5 - z2)^2 + (z5 - z8)^2} \tag{2.36}$$

$$G2 = \sqrt{(z5 - z4)^2 + (z5 - z6)^2} \tag{2.37}$$

$$G3 = \sqrt{(z5 - z1)^2 + (z5 - z9)^2} \tag{2.38}$$

Fig. 2.8 Matrix of 3 × 3 of
the index Z_i, that is
indicating the calculation of
the gradient in the four
directions

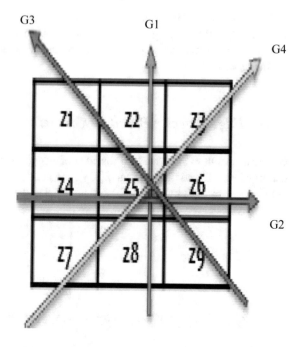

$$G4 = \sqrt{(z5 - z7)^2 + (z5 - z3)^2} \tag{2.39}$$

$$G = G1 + G2 + G3 + G4 \tag{2.40}$$

2.3.1.1 FOM Metrics

Once edge detection with a particular technique is performed, it is necessary to use
some evaluation method to determine whether the result is good or better than other
existing edge detection methods in digital images.

In the literature, we can find different metrics to evaluate the detected edges, to
perform these evaluations, a measure of differences that is frequently used is the
quadratic average distance proposed by Pratt [28], and it is also called Figure of
Merit (FOM). This measure represents the variation that exists from a real point
(calculated) from the ideal edge.

$$FOM = \frac{1}{\max(I_I, I_A)} \sum_{i=1}^{I_A} \frac{1}{1 + \propto d_i^2} \tag{2.41}$$

where I_A represents the number of detected edge points, I_I is the number of points on the ideal edge, d(i) is the distance between the current pixel and its correct position in the reference image and α is a scaling parameter (normally 1/9). To apply (2.41) we require the synthetic image and its reference. A FOM = 1 corresponds to a perfect match between the edge ideal and the detected edge points.

2.4 Modular Neural Network

The focus is on the aggregation operators that are based on measures, in particular the GIT2SI that is applied to a modular neural network (MNN) for the case of face recognition. MNNs use others techniques to add the information of the modules, like type-1 and type-2 fuzzy systems [29–31], the fuzzy Choquet integral [32], the Sugeno integral [33] a probabilistic sum integrator [34], a novel Bayesian learning method [35] and self-organizing maps [36]. A list of some of the most used methods in the integration of MNNs are shown in the first column of Table 2.2, and in the second column we can observe some operators used when performing the process of numerical aggregation.

2.4.1 Methodology for Face Recognition

The Sugeno integral has been used in a variety of applications; in this case we use the GT2SI as an aggregation operator for a MNN applied to face recognition. The experiment was performed as follows: to each of the selected databases we applied a preprocessing step by using edge detectors to highlight the characteristics of the

Table 2.2 Combining information sources

Integration methods	Aggregation operators
The winner takes it all	Arithmetic mean
Linear combination of results	Geometric mean
Voting mechanisms	Weighted mean
Models in series	OWA
Discrete logic	OWA weighted
Type-1 and interval Type-2 fuzzy logic	Harmonic mean
Fuzzy Sugeno integral	Choquet integral
Interval Type-2 Fuzzy integral	Sugeno integral
Self-organizing maps	Interval type-2 Sugeno integral
	Bayesian learning method
	Probabilistic sum integrator

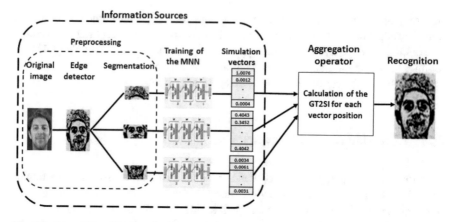

Fig. 2.9 Proposed architecture for face recognition

information. Subsequently, we performed the image division into 3 segments for training with each of the modules of the MNN. Then the cross-validation method was applied for the selection of the training and testing data, finally we performed the aggregation process through the implementation of the GT2SI for achieving the recognition. This procedure can be found in Fig. 2.9.

In this case, we considered face databases for the experiment; however, the proposed aggregation operator can be used in any application where it is necessary to perform the process of numerical aggregation.

The sources of information that should be aggregated are represented by the outputs of each of the MNN modules. Of course, it is necessary to assign a fuzzy density $\mu(\{x_i\})$ to each of the sources of information.

The calculation of fuzzy densities was performed using (2.42) where i, j and k represent the values of the fuzzy densities assigned to each information source. For the experiments, the i, j and k variables were increased by 0.4 in each iteration.

$$\overrightarrow{FD} = \sum_{i=0.1}^{0.9} \sum_{j=0.1}^{0.9} \sum_{k=0.1}^{0.9} [i, j, k] \tag{2.42}$$

In the experiments we performed 27 tests per each simulation with the edge detectors. Each test is the result of the permutations made on i, j and k with the values of 0.1, 0.5 and 0.9, which represent the fuzzy densities assigned to each information source, in this case for each module. The parameters of the fuzzy densities are in this case assigned arbitrarily, however the selected parameters are not optimized, so it is necessary to apply some methodology to find the optimal densities for the particular problem.

2.4.2 Face Databases

To validate the proposed approach, we used two face databases; the first one is the ORL database [37]. This database has faces of 40 people with 10 samples of each individual. All the images of the ORL face database are in a frontal position with a rotation of up to about 20°, for some a tolerance of tilting and against a dark homogeneous background. The images of some subjects were taken at different times and have varying lighting, facial expressions with open and closed eyes, smiling or not smiling and facial details, such as glasses or no glasses. To each of the faces of the database we apply a pre-processing step by using the Morphological Gradient edge detector [38] in order to highlight the main features. The images of the ORL database have a dimension of 112×92. The Cropped Yale database has faces of 38 people with 10 samples of each person, each image has a size of 168×192, and a similar process is applied.

2.4.3 Pre-processing with the Morphological Gradient (MG) Edge Detector

When working with a pattern recognition system is necessary to keep as much information as possible, so that making use of an edge detector method we have that the resulting images retain more detail than the original images. We can define an edge detection method as a procedure that consists on identifying variations that exists in the light intensity, which can be used as a factor to determine properties or characteristics of the elements present in the faces.

Different researchers have dealt with the problem developing edge detectors, in [38, 39] has been shown that performing a pre-processing to the images, using an edge detector, and significantly improves the recognition rate. For this reason, in this case the Type-1 and interval Type-2 Morphological Gradient edge detectors are used for each one of the face databases. In Figs. 2.10 and 2.11, in (a) we can appreciate the original images, in (b) we can find the images after applying the Type-1 Morphological Gradient and in (c) we can observe the images after applying the interval Type-2 MG.

2.4.4 Cross Validation Method

In the realization of the experiments we established the quantity of persons as p and s as the number of samples per person. For each of the databases, tests were made using the k-fold cross validation method, with $k = 5$.

(a)

(b)

(c)

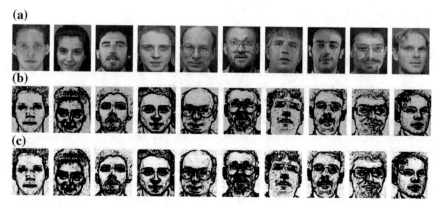

Fig. 2.10 **a** Original faces of the ORL database, **b** image after applying MGT1FLS, **c** image after applying MGIT2FLS

(a)

(b)

(c)

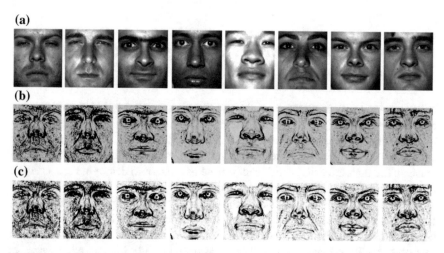

Fig. 2.11 **a** Original faces of the Cropped Yale database, **b** images after applying the MGT1FLS method, **c** images after applying the MGIT2FLS method

We can generalize the calculation of the fold size m or number of samples in each fold, dividing the total number of samples s for each person between the folds number, and then multiplying the result by the people number p (2.43). The training data set size i (2.44) can be calculated as the number of samples in $k - 1$ folds m, and the test data set size t (2.45) is the number of samples in only one fold [40].

Table 2.3 Distribution of the data in the folds

Database	People quantity (p)	Samples per person (s)	Fold size (m)
ORL	40	10	80
Cropped Yale	38	10	76

$$m = (s/k) * p \tag{2.43}$$

$$i = m * (k - 1) \tag{2.44}$$

$$t = m \tag{2.45}$$

The total quantity of samples used for each person has value of ten for both the ORL and the Cropped Yale databases; then if the size m of each five-fold is two, the quantity of samples for the training of each people is eight and for testing is two.

The total number of samples used for each person are of eight for the Cropped Yale databases; then if the size m of each five-fold is two, the number of samples for the training for each person is eight and for testing two, as can be noted in Table 2.3.

2.4.5   Training of the MNN

The concept of artificial neural networks (ANN) was introduced by W. S. McCullough and W. Pitts in 1943 [41] inspired by the biological neural networks of the human brain. An ANN is defined as an information processing system that has some performance characteristics similar to the biological neurons. The ANNs are considered mathematical models of the neural biology or the human cognition [42]. We applied the neural network concept due to the ability to learn from data, yet, there are engineering problems or applications that cannot be computed on a single neural network due of its difficulty or the amount information that is handled. In these cases, we can divide the problem into sub-problems or subtasks that are less complex and use a MNN so that each module is able to solve a part of the problem, thus reducing the corresponding complexity [43–45]. So it is necessary to apply an aggregation operator that allows to integrate the information from the different sources (modules of the MNN) to give the total solution to the problem. We trained a modular neural network of 3 modules with each face database. To each of the images the MGT1FLS and MGIT2FLS edge detectors are applied, then each image is split into three horizontal sections. Each section of the image is used as training data for the MNN, and after that we proceed to perform the integration of information. This methodology is described in more detail in Fig. 2.9.

Table 2.4 Parameters of the MNN

Architecture	Parameters
Training method	Traingdx
Hidden layers	2
Neurons per layer	200
Epochs	500
Error goal	0.0001

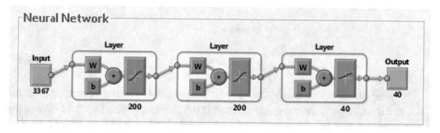

Fig. 2.12 Architecture of each monolithic neural network using the ORL database

In Table 2.4 we can find the training parameters assigned to each monolithic neural network of the MNN, which were proposed in [46], and in Fig. 2.12 we can appreciate the architecture of a monolithic neural network of the MNN.

The selection of training data was performed using the cross-validation method, which goal is to ensure that the results obtained are independent of the partition used to select both the training and the test data.

2.4.6 Experiment

The work consists on obtaining a data set of the ORL and Cropped Yale benchmark face databases by applying each of the edge detectors, and then train a MNN to compare the percentages of recognition applying the k-fold cross validation method [47]. The sequence of steps performed can be found in the diagram of Fig. 2.13.

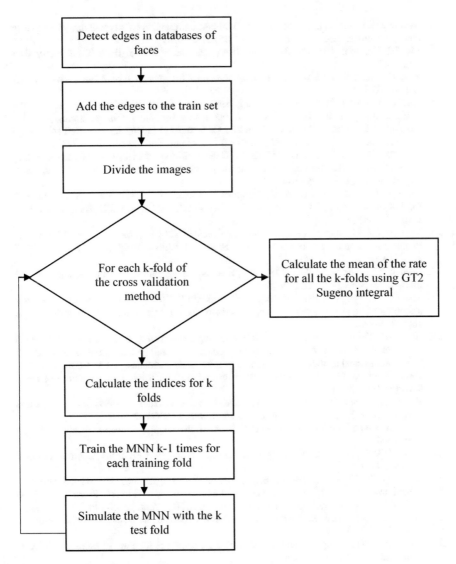

Fig. 2.13 Procedure for recognition

References

1. Zadeh LA (1965) Fuzzy sets. Inf Control 8:338–353
2. Sugeno M (1974) Theory of fuzzy integrals and its applications, Doctoral Thesis, Tokyo Institute of Technology
3. Mendel J (2000) Uncertain. Rule-based fuzzy logic systems. Prentice Hall PTR, pp 555
4. Liu F (2008) An efficient centroid type-reduction strategy for general type-2 fuzzy logic system. Inf Sci 178(9):2224–2236

5. Mendel JM, Fellow L, Liu F, Zhai D (2009) α-plane representation for type-2 fuzzy sets: theory and applications. IEEE Trans Fuzzy Syst 17(5):1189–1207
6. Mendel JM, John RIB (2002) Type-2 fuzzy sets made simple. IEEE Trans Fuzzy Syst 10(2):117–127
7. Zhai D, Mendel JM (2011) Uncertainty measures for general type-2 fuzzy sets. Inf Sci 181(3):503–518
8. Zhai D, Mendel J (2010) Centroid of a general type-2 fuzzy set computed by means of the centroid-flow algorithm. In: Proceedings of international conference fuzzy systems, pp 1–8
9. González CI, Melin P, Castro JR, Mendoza O, Castillo O (2016) An improved sobel edge detection method based on generalized type-2 fuzzy logic. Soft Comput 20(2):773–784
10. Sanchez MA, Castillo O, Castro JR (2015) Generalized type-2 fuzzy systems for controlling a mobile robot and a performance comparison with Interval type-2 and type-1 fuzzy systems. Expert Syst Appl 42(14):5904–5914
11. Melin P, González CI, Castro JR, Mendoza O, Castillo O (2014) Edge-detection method for image processing based on generalized type-2 fuzzy logic. IEEE Trans Fuzzy Syst 22(6):1515–1525
12. Martínez GE, Mendoza O, Castro JR, Melin P, Castillo O (2013) Generalized type-2 fuzzy logic in response integration of modular neural networks. In: IFSA World Congress and NAFIPS Annual Meeting (IFSA/NAFIPS)
13. Bezdek JC, Keller J, Pal NR (2005) Fuzzy models and algorithms for pattern recognition and image processing. Springer, New York
14. Mendez-Vazquez A, Gader P, Keller JM, Chamberlin K (2008) Minimum classification error training for Choquet integrals with applications to landmine detection. IEEE Trans Fuzzy Syst 16(1):225–238
15. Huang KK, Shieh J, Lee KJ, Wu SN (2010) Applying a generalized Choquet integral with signed fuzzy measure based on the complexity to evaluate the overall satisfaction of the patients, vol 5. In: International conference on machine learning and cybernetics, pp 2377–2382
16. Torra V, Narukawa Y (2007) Modeling decisions, information fusion and aggregation operators. Springer, Heidelberg
17. Verikas A, Lipnickas A, Malmqvist K, Bacauskiene M, Gelzinis A (1999) Soft combination of neural classifiers: a comparative study. Pattern Recogn 20(4):429–444
18. Timonin Mikhail (2013) Robust optimization of the Choquet integral. Fuzzy Sets Syst 213:27–46
19. Canny J (1986) A computational approach to edge detection. IEEE Trans Pattern Anal Mach Intell 8(2):679–698
20. Sobel I (1970) Camera models and perception, Ph.D. thesis, Stanford University, Stanford
21. Prewitt JMS (1970) Object enhancement and extraction. In: Lipkin BS, Rosenfeld A (eds) Picture analysis and psychopictorics. Academic Press, New York, pp 75–149
22. Roberts L (1965) Machine perception of 3-D solids, optical and electro-optical information processing. MIT Press
23. Kirsch R (1971) Computer determination of the constituent structure of biological images. Comput Biomed Res 4:315–328
24. Hua L, Cheng H, Zhanga M (2007) A high performance edge detector base d on fuzzy inference rules. J Inf Sci 177(21):4768–4784 (Elsevier Science)
25. Mendoza O, Melin P, Licea G (2007) A new method for edge detection in image processing using interval type-2 fuzzy logic. In: IEEE international conference on granular computing (GRC 2007). Silicon Valley, CA, USA
26. Melin P, Mendoza O, Castillo O (2010) An improved method for edge detection based on interval type-2 fuzzy logic. Expert Syst Appl 37:8527–8535
27. Mendoza O, Melin P, Licea G (2009) Interval type-2 fuzzy logic for edge detection in digital images. Int J Intell Syst 24(11):1115–1134
28. Pratt WK (1978) Digital image processing. Wiley, New York
29. Hidalgo D (2008) Fuzzy inference systems type-1 and type-2 as integration methods in neural networks for multimodal biometrics and me-optimization by means of genetic algorithms, Master Thesis, Tijuana Institute of Technology

30. Sánchez D, Melin P (2010) Modular neural network with fuzzy integration and its optimization using genetic algorithms for human recognition based on iris, ear and voice biometrics. Soft Comput Recognit Based Biometrics 85–102
31. Sánchez D, Melin P, Castillo O, Valdez F (2012) Modular neural networks optimization with hierarchical genetic algorithms with fuzzy response integration for pattern recognition. MICAI, pp 247–258
32. Kwak KC, Pedrycz W (2005) Face recognition: a study in information fusion using fuzzy Integral. Pattern Recognit Lett 26:719–733
33. Mendoza O, Melin P (2008) Extension of the Sugeno integral with interval type-2 fuzzy logic. In: Fuzzy information processing society, NAFIPS
34. Meena YK, Arya KV, Kala R (2010) Classification using redundant mapping in modular neural networks. Nat Biologically Inspired Comput 554–559
35. Wang P, Xu L, Zhou SM, Fan Z, Li Y, Feng S (2010) A novel Bayesian learning method for information aggregation in modular neural networks. Expert Syst Appl 37(2):1071–1074
36. Horiuchi T, Kato S (2009) A study on Japanese historical character recognition using modular neural networks. In: IEEE fourth international conference on innovative computing, information and control, pp 1507–1510
37. Database ORL Face (2012) Cambridge University Computer Laboratory. http://www.cl.cam.ac.uk/research/dtg/attarchive/facedatabase.html
38. Mendoza O, Melin P, Castillo O, Castro J (2010) Comparison of fuzzy edge detectors based on the image recognition rate as performance index calculated with neural networks. In: Soft computing for recognition based on biometrics, studies in computational intelligence, vol 312. Springer, Berlin, Heidelberg, pp 389–399
39. Mendoza O, Melin P (2011) Quantitative evaluation of fuzzy edge detectors applied to neural networks or image recognition. Advances in Research & Developments in Digital Systems, pp 324–335. In: Stochastic algorithms: foundations and applications. Lecture notes computer science, vol 5792, pp 169–178
40. Melin P, Castillo O, Kacprzyk J (2015) Design of intelligent systems based on fuzzy logic, neural networks and nature-inspired optimization. Springer Switzerland, pp 59–70
41. McCulloh WS, Pitts W (1943) A logical calculus of the ideas immanent in nervous activity. Bull Math Biophys 5:115–133
42. Fausett L (1994) Fundamentals of neural networks. Prentice Hall, New Jersey
43. Bo YC, Qiao JF, Yang G (2011) A modular neural networks ensembling method based on fuzzy decision-making. IEEE Electr Inf Control Eng 1030–1034
44. Iryna T, Kochan V, Sachenko A (2006) Recognition of multi-sensor output signal using modular neural networks approach. In: Modern problems of radio engineering, telecommunications, and computer science, pp 155–158
45. Liu Y, Yao X (1997) Evolving modular neural networks which generalise well. IEEE Evol Comput 605–610
46. Vazirani H, Kala R, Shukla A, Tiwari R (2010) Diagnosis of breast cancer by modular neural network . IEEE Int J Biomed Eng Technol 194–211
47. Melin P, Felix C, Castillo O (2005) Face recognition using modular neural networks and the fuzzy Sugeno integral for response integration. Int J Intell Syst 20(2):275–290

Chapter 3
Proposed Method for the Type-2 Fuzzy Sugeno Integral

3.1 Generalized Interval Type-2 Sugeno Integral

The main goal of this work is to extend to the Sugeno integral using generalized type-2 fuzzy systems. As a consequence, the interval type-2 Sugeno integral and the interval type-2 Choquet integral were developed to make comparisons with the proposed method [1–7].

The series of steps used to perform the extension are presented below:

(1) First, depending on the problem, we need to define the number n of information sources, the information sources are denoted by $D(x_i)$, and the fuzzy densities assigned to each information source are represented by $M(x_i) \in (0, 1)$.

(2) To obtain a GT2SI it is necessary to evaluate each $D(x_i)$ and $M(x_i)$ with a generalized type-2 membership function, in this case with a triangular in the primary and a Gaussian in the secondary membership, using Eq. (2.12). The function is represented by $\tilde{\mu}(x, u) = trigausstype2(x, u, [a_1, b_1, c_1, a_2, b_2, c_2, \rho])$, and we can appreciate the plot of the membership function in Fig. 3.1.

(3) After this, it is necessary to calculate the α_i cuts for each $M(x_i)$ and $D(x_i)$ to obtain $\mu(M_{iL\alpha_i}(x_i))$, $\mu(M_{iR\alpha_i}(x_i))$ and $\mu(D_{L\alpha_i}(x_i))$, $\mu(D_{R\alpha_i}(x_i))$.

(4) Based on Eq. (2.17) the next step is to calculate the λ parameter and the α_i cuts for λ_L and λ_R using the Eqs. (3.1) and (3.2):

$$f\left(\lambda_{L\alpha_i}\right) = \left\{ \prod_{i=1}^{n} \left(1 + M_{iL\alpha_i}(x_i)\lambda_{L\alpha_i}\right) \right\} - \left(1 + \lambda_{L\alpha_i}\right) = 0 \qquad (3.1)$$

$$f\left(\lambda_{R\alpha_i}\right) = \left\{ \prod_{i=1}^{n} \left(1 + M_{iR\alpha_i}(x_i)\lambda_{R\alpha_i}\right) \right\} - \left(1 + \lambda_{R\alpha_i}\right) = 0 \qquad (3.2)$$

© The Author(s), under exclusive license to Springer Nature Switzerland AG 2020
P. Melin and G. E. Martinez, *Extension of the Fuzzy Sugeno Integral Based on Generalized Type-2 Fuzzy Logic*, SpringerBriefs in Computational Intelligence, https://doi.org/10.1007/978-3-030-16416-4_3

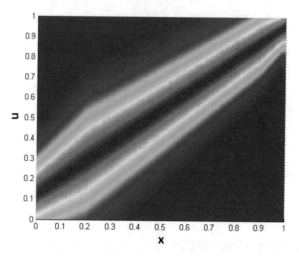

Fig. 3.1 Generalized membership function trigausstype-2

once we obtain $\lambda_{L\alpha_i}$ and $\lambda_{R\alpha_i}$ we now have to calculate the fuzzy measures $\mu_{L\alpha_i}(A_i)$ and $\mu_{R\alpha_i}(A_i)$ with (2.33)–(2.36) by extending Eq. (2.18).

$$\mu_{L\alpha_i}(A_1) = \mu_{L\alpha_i}(x_1) \tag{3.3}$$

$$\mu_{L\alpha_i}(A_i) = \mu_{L\alpha_i}(x_i) + \mu_{L\alpha_i}(A_{i-1}) + \lambda_{L\alpha_i}\mu_{L\alpha_i}(x_i)\mu_{L\alpha_i}(A_{i-1}) \tag{3.4}$$

$$\mu_{R\alpha_i}(A_1) = \mu_{R\alpha_i}(x_1) \tag{3.5}$$

$$\mu_{R\alpha_i}(A_i) = \mu_{R\alpha_i}(x_i) + \mu_{R\alpha_i}(A_{i-1}) + \lambda_{R\alpha_i}\mu_{R\alpha_i}(x_i)\mu_{R\alpha_i}(A_{i-1}) \tag{3.6}$$

(5) Next, based on Eq. (2.20) we need calculate the Sugeno integral for each α_i with Eq. (3.7).

$$h(\tilde{\sigma}_{\alpha_1}, \tilde{\sigma}_{\alpha_2}, \ldots, \tilde{\sigma}_{\alpha_n}) = \sqcup_{l=1}^{n}\left(\sqcap[h_{L\alpha_1}, h_{R\alpha_1}], \sqcap[h_{L\alpha_2}, h_{R\alpha_2}], \ldots, \sqcap[h_{L\alpha_n}, h_{R\alpha_n}]\right) \tag{3.7}$$

where each $\tilde{\sigma}_{\alpha_i}$ is an interval of the form (3.8)

$$\tilde{\sigma}_{\alpha_i} = \sqcup_{i=1}^{n}\left(\sqcap\left([\mu(D_{L\alpha_i}(x_i)), \mu_{L\alpha_i}(A_i)], [\mu(D_{R\alpha_i}(x_i)), \mu_{R\alpha_i}(A_i)]\right)\right) \tag{3.8}$$

(6) Finally, using Eq. (3.9) we calculate the supremum of the $\tilde{\sigma}_{\alpha_i}$ to obtain the GT2SI.

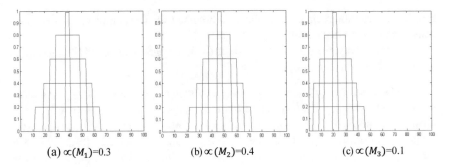

(a) $\propto(M_1)$=0.3   (b) $\propto(M_2)$=0.4   (c) $\propto(M_3)$=0.1

Fig. 3.2 Alpha cuts of the information sources

$$h = \overset{sup}{\underset{i}{}}\left(h\left(\tilde{\sigma}_{\alpha_i}\right)\right) \tag{3.9}$$

To exemplify the proposed method, the following parameters are defined below, if we have the information sources given as $\sigma(x_1) = 0.9$, $\sigma(x_2) = 0.6$, $\sigma(x_3) = 0.3$ and the fuzzy densities defined as $\mu c(x_1) = 0.3$, $\mu c(x_2) = 0.4$ and $\mu c(x_3) = 0.1$, associated to each $\sigma(x_i)$, the first step is evaluate each $\sigma(x_i)$ with the membership function *trigausstype2(x, u, [−0.5 0.9 1.3 0.2 1.2 1.2 0.1])*, to generate the alpha cuts of the fuzzy densities, The representation of the membership function can be found in Fig. 3.2. For each of the α_i cuts we obtain a left interval and right interval of uncertainty.

The parameters of the membership function were arbitrarily designated and were manually adapted, taking care that when evaluating the information sources and the fuzzy densities assigned to each source, the obtained values comply with the initial conditions defined by Sugeno. Then, using the parameters previously defined, we calculate the α cuts in the points 0.2, 0.4, 0.6, 0.8 and 0.99 for each fuzzy density $M(x_i)$ to obtain $\mu(M_{iL\alpha i}(x_i))$ and $\mu(M_{iR\alpha i}(x_i))$:

$\mu(M_{1L\alpha 0.2}) = 0.1064$, $\mu(M_{1L\alpha 0.4}) = 0.1723$, $\mu(M_{1L\alpha 0.6}) = 0.2237$, $\mu(M_{1L\alpha 0.8}) = 0.2750$, $\mu(M_{1L\alpha 0.99}) = 0.3538$

$\mu(M_{1R\alpha 0.2}) = 0.6436$, $\mu(M_{1R\alpha 0.4}) = 0.5577$, $\mu(M_{1R\alpha 0.6}) = 0.5263$, $\mu(M_{1R\alpha 0.8}) = 0.4750$, $\mu(M_{1R\alpha 0.99}) = 0.3962$

$\mu(M_{2L\alpha 0.2}) = 0.2060$, $\mu(M_{2L\alpha 0.4}) = 0.2680$, $\mu(M_{2L\alpha 0.6}) = 0.3162$, $\mu(M_{2L\alpha 0.8}) = 0.3644$, $\mu(M_{2L\alpha 0.99}) = 0.4384$

$\mu(M_{2R\alpha 0.2}) = 0.7106$, $\mu(M_{2R\alpha 0.4}) = 0.6487$, $\mu(M_{2R\alpha 0.6}) = 0.6005$, $\mu(M_{2R\alpha 0.8}) = 0.5523$, $\mu(M_{2R\alpha 0.99}) = 0.3962$

$\mu(M_{3L\alpha 0.2}) = 0.0100$, $\mu(M_{3L\alpha 0.4}) = 0.0241$, $\mu(M_{3L\alpha 0.6}) = 0.0708$, $\mu(M_{3L\alpha 0.8}) = 0.1174$, $\mu(M_{3L\alpha 0.99}) = 0.1890$

$\mu(M_{3R\alpha 0.2}) = 0.5425$, $\mu(M_{3R\alpha 0.4}) = 0.3926$, $\mu(M_{3R\alpha 0.6}) = 0.3454$, $\mu(M_{3R\alpha 0.8}) = 0.2992$, $\mu(M_{3R\alpha 0.99}) = 0.2276$

For each data from each information source, of each alpha cut an interval is generated, so if 5 cuts are defined for 3 fuzzy densities assigned to three information sources, after evaluating the information with GT2SI, it results in 30 densities. In

Fig. 3.2 we can appreciate the alpha cuts obtained for the parameter of the fuzzy densities assigned to each information source.

In the same way, we calculate the α_i for the information sources D (x_i) to obtain $\mu(D_{L\alpha i}(x_i))$ and $\mu(D_{R\alpha i}(x_i))$.

$\mu(D_{1L\alpha 0.2}) = 0.7041, \mu(D_{1L\alpha 0.4}) = 0.7460, \mu(D_{1L\alpha 0.6}) = 0.7787, \mu(D_{1L\alpha 0.8}) = 0.8114, \mu(D_{1L\alpha 0.99}) = 0.8615$

$\mu(D_{1R\alpha 0.2}) = 1, \mu(D_{1R\alpha 0.4}) = 1, \mu(D_{1R\alpha 0.6}) = 0.9713, \mu(D_{1R\alpha 0.8}) = 0.9386, \mu(D_{1R\alpha 0.99}) = 0.8885$

$\mu(D_{2L\alpha 0.2}) = 0.4053, \mu(D_{2L\alpha 0.4}) = 0.4592, \mu(D_{2L\alpha 0.6}) = 0.5012, \mu(D_{2L\alpha 0.8}) = 0.5432, \mu(D_{2L\alpha 0.99}) = 0.6067$

$\mu(D_{2R\alpha 0.2}) = 0.8447, \mu(D_{2R\alpha 0.4}) = 0.7908, \mu(D_{2R\alpha 0.6}) = 0.7488, \mu(D_{2R\alpha 0.8}) = 0.7086, \mu(D_{2R\alpha 0.99}) = 0.6424$

$\mu(D_{3L\alpha 0.2}) = 0.1064, \mu(D_{3L\alpha 0.4}) = 0.1723, \mu(D_{3L\alpha 0.6}) = 0.2237, \mu(D_{3L\alpha 0.8}) = 0.2750, \mu(D_{3L\alpha 0.99}) = 0.3538$

$\mu(D_{3R\alpha 0.2}) = 0.6436, \mu(D_{3R\alpha 0.4}) = 0.5777, \mu(D_{3R\alpha 0.6}) = 0.5263, \mu(D_{3R\alpha 0.8}) = 0.4750, \mu(D_{3R\alpha 0.99}) = 0.3962$

Once the alpha cuts of the fuzzy densities $\mu(M_{iL\alpha i}(x_i))$ and $\mu(M_{iR\alpha i}(x_i))$ are calculated, is necessary to estimate the $\lambda_{L\alpha i}$ and $\lambda_{R\alpha i}$, by Eqs. (3.1 and 3.2).

$$\lambda_{L\alpha 0.2} = 22.5836, \lambda_{L\alpha 0.4} = 8.1333, \lambda_{L\alpha 0.6} = 3.1251, \lambda_{L\alpha 0.8} = 1.2779, \lambda_{L\alpha 0.99} = 0.0612$$
$$\lambda_{R\alpha 0.2} = -0.9162, \lambda_{R\alpha 0.4} = -0.8458, \lambda_{R\alpha 0.6} = -0.7589, \lambda_{R\alpha 0.8} = -0.6274, \lambda_{R\alpha 0.99} = -0.2710$$

Then using Eqs. (3.3–3.6) the next step is calculate the fuzzy measures $\mu_{L\alpha i}(A_i)$ and $\mu_{R\alpha i}(A_i)$ for each α_i

$\mu_{L\alpha 0.2}(A_1) = 0.1064, \mu_{L\alpha 0.4}(A_1) = 0.1723, \mu_{L\alpha 0.6}(A_1) = 0.2237, \mu_{L\alpha 0.8}(A_1) = 0.2750, \mu_{L\alpha 0.99}(A_1) = 0.3538$

$\mu_{R\alpha 0.2}(A_1) = 0.6436, \mu_{R\alpha 0.4}(A_1) = 0.5777, \mu_{R\alpha 0.6}(A_1) = 0.5263, \mu_{R\alpha 0.8}(A_1) = 0.4750, \mu_{R\alpha 0.99}(A_1) = 0.3962$

$\mu_{L\alpha 0.2}(A_2) = 0.8076, \mu_{L\alpha 0.4}(A_2) = 0.8159, \mu_{L\alpha 0.6}(A_2) = 0.7609, \mu_{L\alpha 0.8}(A_2) = 0.7674, \mu_{L\alpha 0.99}(A_2) = 0.8017$

$\mu_{R\alpha 0.2}(A_2) = 0.9352, \mu_{R\alpha 0.4}(A_2) = 0.9094, \mu_{R\alpha 0.6}(A_2) = 0.8869, \mu_{R\alpha 0.8}(A_2) = 0.8627, \mu_{R\alpha=0.99}(A_2) = 0.8231$

$\mu_{L\alpha 0.2}(A_3) = 1, \mu_{L\alpha=0.4}(A_3) = 1, \mu_{L\alpha=0.6}(A_3) = 1, \mu_{L\alpha=0.8}(A_3) = 1, \mu_{L\alpha=0.99}(A_3) = 1$

$\mu_{R\alpha 0.2}(A_3) = 1, \mu_{R\alpha 0.4}(A_3) = 1, \mu_{R\alpha 0.6}(A_3) = 1, \mu_{R\alpha 0.8}(A_3) = 1, \mu_{R\alpha 0.99}(A_3) = 1$

Next, it is necessary to perform the calculation of the GT2SI for the α_L cut and the α_R cut in 0.2 in the following way:

$$\sigma_{\alpha 0.2} = \sqcup(\sqcap([\mu(D_{L\alpha 0.2}(x_1)), \mu_{L\alpha 0.2}(A_1)], [\mu(D_{R\alpha 0.2}(x_1)), \mu_{R\alpha 0.2}(A_1)]),$$
$$\sqcap([\mu(D_{L\alpha 0.2}(x_2)), \mu_{L\alpha 0.2}(A_2)], [\mu(D_{R\alpha 0.2}(x_2)), \mu_{R\alpha 0.2}(A_2)]), \sqcap([\mu(D_{L\alpha 0.2}(x_3)), \mu_{L\alpha 0.2}(A_3)], [\mu(D_{R\alpha 0.2}(x_3)), \mu_{R\alpha 0.2}(A_3)]))$$

The values obtained above are then replaced to carry out the following calculation:

$$\sigma_{\alpha 0.2} = \sqcup(\sqcap([0.7441, 0.1064], [1, 0.6436]),$$
$$\sqcap([0.4053, 0.8076], [0.8447, 0.9352]), \sqcap([0.1064, 1], [0.6436, 1]))$$
$$\sigma_{\alpha 0.2} = \sqcup([0.1064, 0.6436], [0.4053, 0.8447], [0.1064, 0.64336])$$
$$\sigma_{\alpha 0.2} = [0.4053, 0.8447]$$

In the same way, we perform the calculation for the α_L cut and the α_R cut in 0.4:

$$\sigma_{\alpha 0.4} = \sqcup(\sqcap([0.7460, 0.1723], [1, 0.5777]),$$
$$\sqcap([0.4592, 0.8159], [0.7908, 0.9094]), \sqcap([0.1723, 1], [0.5777, 1]))$$
$$\sigma_{\alpha 0.4} = \sqcup([0.1723, 0.5777], [0.4592, 0.7908], [0.1723, 0.5777])$$
$$\sigma_{\alpha 0.4} = [0.4592, 0.7908]$$

The result of calculating the α_L cut and the α_R cut in 0.6 is:

$$\sigma_{\alpha 0.6} = \sqcup(\sqcap([0.7787, 0.2237], [0.9713, 0.5263]),$$
$$\sqcap([0.5012, 0.7609], [0.7488, 0.8869]), \sqcap([0.2237, 1], [0.5263, 1]))$$
$$\sigma_{\alpha 0.6} = \sqcup([0.2237, 0.5263], [0.5012, 0.7488], [0.2237, 0.5263])$$
$$\sigma_{\alpha 0.6} = [0.5012, 0.7488]$$

For the α_L cut and the α_R cut in 0.8 the calculation is performed as follows:

$$\sigma_{\alpha 0.8} = \sqcup(\sqcap([0.8114, 0.2750], [0.9386, 0.4750]),$$
$$\sqcap([0.5432, 0.7674], [0.7086, 0.8627]), \sqcap([0.2750, 1], [0.4750, 1]))$$
$$\sigma_{\alpha 0.8} = \sqcup([0.2750, 0.4750], [0.5432, 0.7086], [0.2750, 0.4750])$$
$$\sigma_{\alpha 0.8} = [0.5432, 0.7086]$$

In the cut 0.99 the calculation of α_L and α_R is carried out as follows:

$$\sigma_{\alpha 0.99} = \sqcup(\sqcap([0.8615, 0.3538], [0.8885, 0.3962]),$$
$$\sqcap([0.6076, 0.8017], [0.6424, 0.8231]), \sqcap([0.3538, 1], [0.3962, 1]))$$
$$\sigma_{\alpha 0.99} = \sqcup([0.3538, 0.3962], [0.6076, 0.6424], [0.3538, 0.3962])$$
$$\sigma_{\alpha 0.99} = [0.6076, 0.6424]$$

Once the SI has been calculated for all the α_i cuts, we use (3.9) to obtain an approximation of the GT2SI. Using the data assigned in the previous example, we calculated the SI and the IT2SI with the purpose of comparing the behavior of the integrals. We can find the results obtained for the three methods in Table 3.1. It can be observed that by applying Eq. (2.20), that corresponds to the traditional SI a result of 0.6 was obtained, while using the IT2SI, the uncertainty interval of [0.5, 0.7] was obtained; In this case, Eq. (2.32) was used to calculate an approximation in order

Table 3.1 Results obtained using SI, IT2SI and GT2SI

Method	Interval obtained	Result
SI	–	0.6
IT2SI	[0.5, 0.7]	0.6
GT2SI	[0.4053, 0.8447] [0.4592, 0.7908] [0.5012, 0.7488] [0.5432, 0.7068] [0.6076, 0.6424]	0.625

to obtain a numerical value representing the output of the IT2SI at the moment of taking it to an actual application, the IT2SI obtained was of 0.6. With the GT2SI, an uncertainty interval was obtained for each of the α cuts, in this case five α cuts were defined, and so we have as output five intervals (one for each α cut), using Eq. (3.9), the GT2SI was of 0.625.

When working with generalized type-2 fuzzy systems, the challenge is to identify the number of alpha cuts to be performed according to the application, as well as the position of each cut. In this work to approximate the GT2 FSs the α-planes theory was used (described in Sect. 2.1.3.1); for this reason, we considered varying the number of α-planes necessary to approximate the output. For this example, a sample was obtained by varying both the number of cuts as well as the position of each one. In Table 3.2 a summary of the obtained results is presented. Were conducted 14 tests arbitrarily; It is possible to appreciate that the best result was obtained both with tests 1 and 3 with 5 and 3 alpha cuts respectively and the obtained FOM value was of 0.625. However tests 6 and 8 were also performed using 5 and 3 cuts each, but the results did not were so satisfactory, getting a FOM of 0.6143 and 0. 6186 respectively, this is because the positions of the cuts made in each test were different, so it must be taken into account that both the number of cuts and the position play an important role when working with a Generalized type -2 Fuzzy system.

We envision applying the proposed method in applications that have been previously addressed by using simple type-2 systems (like in [8–12]) and we expect that results can be improved by using the Sugeno Integral.

Table 3.2 Results of the variation of number of cuts and the position in a GT2SI

Test	Result using GT2SI	Number of ∝ cuts	Position of the ∝ cuts											
1	0.625	5			[0.2		0.4		0.6		0.8		0.99	1]
2	0.6196	6	[0		0.2		0.4		0.6		0.8		0.99	1]
3	0.625	3				[0.3			0.6			0.9]		
4	0.5929	1	[0]											
5	0.6089	2	[0	0.1]										
6	0.6143	3	[0	0.1	0.2]									
7	0.617	4	[0	0.1	0.2	0.3]								
8	0.6186	5	[0	0.1	0.2	0.3	0.4]							
9	0.6196	6	[0	0.1	0.2	0.3	0.4	0.5]						
10	0.6204	7	[0	0.1	0.2	0.3	0.4	0.5	0.6]					
11	0.621	8	[0	0.1	0.2	0.3	0.4	0.5	0.6	0.7]				
12	0.6214	9	[0	0.1	0.2	0.3	0.4	0.5	0.6	0.7	0.8]			
13	0.6218	10	[0	0.1	0.2	0.3	0.4	0.5	0.6	0.7	0.8	0.9]		
14	0.6221	11	[0	0.1	0.2	0.3	0.4	0.5	0.6	0.7	0.8	0.9	0.99]	

References

1. Martínez GE, Mendoza O, Castro JR, Melin P, Castillo O (2013) Generalized type-2 fuzzy logic in response integration of modular neural networks. IFSA World Congress and NAFIPS annual meeting (IFSA/NAFIPS)
2. Martinez GE, Mendoza O, Melin P, Gaxiola F (2016) Comparison between Choquet and Sugeno integrals as aggregation operators for modular neural networks. FUZZ-IEEE 2331–2336
3. Martínez GE, Mendoza O, Castro JR, Melin P, Castillo O (2017) Choquet integral and interval type-2 fuzzy choquet integral for edge detection. In: Nature-inspired design of hybrid intelligent systems, pp 79–97
4. Martínez GE, Mendoza O, Castro JR, Melin P, Castillo O (2016) Comparison between Choquet and Sugeno integrals as aggregation operators for pattern recognition. NAFIPS, pp 1–6
5. Martinez GE, Mendoza O, Castro JR, Rodriguez-Diaz A, Melin P, Oscar Castillo. (2015) Response integration in modular neural networks using Choquet Integral with Interval type 2 Sugeno measures. NAFIPS/WConSC, pp 1–6
6. Martínez GE, Melin P, Mendoza OD, Castillo O (2015) Face recognition with a Sobel edge detector and the Choquet integral as integration method in a modular neural networks. Design of intelligent systems based on fuzzy logic, neural networks and nature-inspired optimization, pp 59–70
7. Martínez GE, Mendoza O, Castro JR, Melin P, Castillo O (2014) Choquet integral with interval type 2 Sugeno measures as an integration method for modular neural networks. WCSC, pp 71–86
8. Melin P, Mancilla A, Lopez M, Mendoza O (2007) A hybrid modular neural network architecture with fuzzy Sugeno integration for time series forecasting. Appl Soft Comput 7(4):1217–1226
9. Melin P, González CI, Castro JR, Mendoza O, Castillo O (2014) Edge-detection method for image processing based on generalized type-2 fuzzy logic. IEEE Trans Fuzzy Syst 22(6):1515–1525
10. González CI, Melin P, Castro JR, Castillo O, Mendoza O (2016) Optimization of interval type-2 fuzzy systems for image edge detection. Appl Soft Comput 47:631–643
11. González CI, Melin P, Castro JR, Mendoza O, Castillo O (2016) An improved sobel edge detection method based on generalized type-2 fuzzy logic. Soft Comput 20(2):773–784
12. Ontiveros E, Melin P, Castillo O (2018) High order α-planes integration: a new approach to computational cost reduction of general type-2 fuzzy systems. Eng Appl of AI 74:186–197

Chapter 4
Simulation Results of the Type-2 Fuzzy Sugeno Integral

4.1 Morphological Gradient Edge Detector Using GT2SI (MG + GT2SI)

The main goal of this work is to use the GT2SI, as an aggregation method in MG edge detectors; therefore, the aggregation done in MG using (2.41) is replaced by the method proposed in Chapter 3. Previous works on edge detectors can be found in [1–4].

In this section, the simulation results of the proposed method applied to synthetic images (as in Fig. 4.1) and real images are presented. To carry out the aggregation process we use G_i defined in the form shown in Fig. 2.8 using Eqs. (2.36–2.39), as the information sources which must be aggregated. In this case instead of using the traditional method of MG proposed in Sect. 2.3.1 to detect the edges in the images, the aggregation process performed in MG by Eq. (2.40), is now replaced by the GT2SI method proposed in Section three to detect the edges in the images.

The aggregation process of the MG + GT2SI is illustrated in Fig. 4.2. As we can note, we first obtain the G_i gradients in the four directions, which are considered the sources of information to which the aggregation method will be applied. As a second step we have to assign the fuzzy densities to each source of information, then the process of aggregation of the gradients is carried out by means of GT2SI to finally we obtain the edge detected in the image. This is based on previous works, such as in [5–12].

Visually it is difficult to appreciate the difference between the detected edges with respect at others methods and for this reason, the method is applied on synthetic images with the Pratt's Figure of Merit (FOM), which was described in Sect. 2.3.1.1 and was defined by Eq. (2.41). This metric is used as performance index to determine the quality of the resulting image. To evaluate the quality of the detected edge using FOM, it is necessary to have a reference image as in Fig. 4.3, which represents an image with the ideal edges.

© The Author(s), under exclusive license to Springer Nature Switzerland AG 2020
P. Melin and G. E. Martinez, *Extension of the Fuzzy Sugeno Integral Based on Generalized Type-2 Fuzzy Logic*, SpringerBriefs in Computational Intelligence, https://doi.org/10.1007/978-3-030-16416-4_4

Fig. 4.1 Sphere synthetic image

To evaluate the quality of the detected edge using the FOM (2.41) we require the synthetic image (Fig. 4.1) and its reference (Fig. 4.3). As a result of the evaluation, a FOM = 1 corresponds to a perfect match between the ideal edge and the detected edge points, and this means that the detected edge is very similar or the same to the ideal edge (Fig. 4.4).

Based on the results of Table 3.2 it was determined to use five α planes with the MG + GT2SI edge detector which were defined in the points $(0.2, 0.4, 0.6, 0.8, 0.99)$ respectively. The calculated gradients determine each of the sources of information to which a fuzzy density should be assigned. In this case, the fuzzy density of 0.6 was arbitrarily defined to each source of information.

In Table 4.1 we can find the results of applying edge detectors at synthetic images. In the first column we can appreciate the image after applying the MG edge detector, the second column is the result of applying MG + SI. The third column corresponds to the images to which MG + IT2SI was applied, and finally, the last column contains the images obtained with MG + GT2SI.

Table 4.2 shows the results obtained after applying the MG, MG + SI, MG + IT2SI and MG + GT2SI edge detectors as the aggregation operators in real images.

We can appreciate that with the FOM values achieved in the Table 4.3, we can determine that the combination of aggregation operators SI, IT2SI and GT2SI with the MG edge detection methods present better result that the traditional MG. The results obtained with the MG + GT2SI are better than the ones calculated by MG + IT2SI and that MG + SI. Also is necessary to highlight that the combination of the aggregation operator SI with techniques of fuzzy logic, improves operator performance with respect to the traditional MG edge detection method.

Fig. 4.2 Diagram that represent the integration of gradients

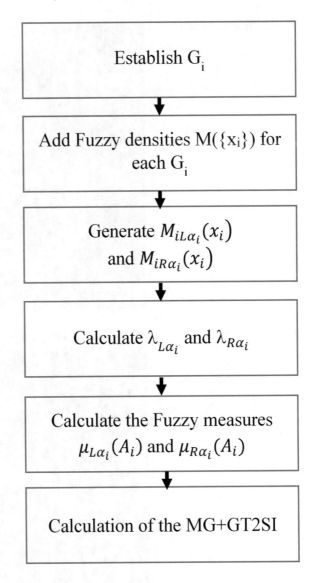

4.2   GT2SI in Modular Neural Networks (GT2SI + MNN)

The proposed aggregation operator (GT2SI) can be used to integrate information sources of different types. In this case the information that is combined is given by the simulation outputs of the modules trained to recognize a different part of the image [13].

In this case study the GT2SI is used, as an aggregation method of the MNN [14]. The experiment consists on applying the well-known benchmark data base of

Fig. 4.3 Sphere reference
image

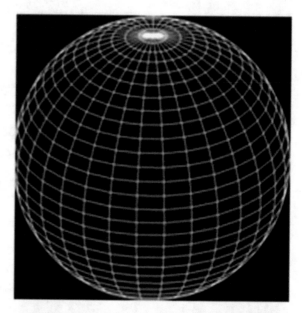

Fig. 4.4 MG edge detected
of the sphere image

Table 4.1 Synthetic images after apply edge detectors

MG	MG + SI	MG + IT2SI	MG + GT2SI

Table 4.2 Real images after applying edge detectors

MG	MG + SI	MG + IT2SI	MG + GT2SI

Table 4.3 FOM obtained after applying edge detectors in synthetic images

Image	FOM			
	GM	MG + SI	MG + IT2SI	MG + GT2SI
	0.8199	0.9408	0.9465	0.9503
	0.8744	0.9477	0.9503	0.9582

images, like ORL (Fig. 2.10) and Cropped Yale (Fig. 2.11) faces databases and then perform training of a modular neural network to compare the recognition rate using the k-fold cross validation method described in the Sect. 2.4.4, which can be found in the Fig. 2.13. Once that the MNN was trained, the proposed GT2SI method is used (Chap. 3) as an integration of responses method for the MNN.

To each of the images of the ORL and Cropped Yale databases, the Morphological gradient type-1 fuzzy logic system (T1MG) and the Morphological gradient type-2 fuzzy logic system (IT2MG) edge detectors are applied. After that a MNN of 3 modules is trained with 80% of the data of each database and the remaining 20% was used for testing. In Table 2.3 we can observe the distribution of the data in the folds.

Once the edge detector is applied, each image is divided into 3 horizontal sections and each of which was used as training data in each of the modules. In Fig. 2.9 we can appreciate the structure of the MNN. In Fig. 2.12 we can observe the architecture of each monolithic neural network. In Table 4.4 we can find the training parameters assigned to each monolithic neural network of the MNN.

The final decision is made using the GT2SI, which is considered to combine the simulation vectors of the MNN to obtain a simple vector. For this study case, five α planes are used with the MNN + GT2SI, which were defined in the points (0.2, 0.4, 0.6, 0.8, 0.99) respectively. The modules of the MNN determine each of the information sources to which a fuzzy density should be assigned. The values 0.1, 0.5 and 0.9 are arbitrarily chosen and all possible permutations were made with this values using Eq. (2.42) for obtain a total of 27 tests for each simulation, as can observed in Table 4.4.

Table 4.4 Fuzzy densities

Test	Fuzzy densities		
	M_1	M_2	M_3
1	0.1	0.1	0.1
2	0.1	0.1	0.5
3	0.1	0.1	0.9
4	0.1	0.5	0.1
5	0.1	0.5	0.5
6	0.1	0.5	0.9
7	0.1	0.9	0.1
8	0.1	0.9	0.5
9	0.1	0.9	0.9
10	0.5	0.1	0.1
11	0.5	0.1	0.5
12	0.5	0.1	0.9
13	0.5	0.5	0.1
14	0.5	0.5	0.5
15	0.5	0.5	0.9
16	0.5	0.9	0.1
17	0.5	0.9	0.5
18	0.5	0.9	0.9
19	0.9	0.1	0.1
20	0.9	0.1	0.5
21	0.9	0.1	0.9
22	0.9	0.5	0.1
23	0.9	0.5	0.5
24	0.9	0.5	0.9
25	0.9	0.9	0.1
26	0.9	0.9	0.5
27	0.9	0.9	0.9

4.2.1   Simulation Results Using ORL Database

Each presented result is the average of 27 tests performed for each training of the MNN in which variations to the fuzzy densities are made by Eq. (2.42), as can be found in Table 4.5. In this case an average value closer to 1 means that it has a greater percentage of recognition.

Next, we can appreciate the results of the training using the ORL database. The mean of using the GT1MG in T1MG and IT2MG are presented in Tables 4.5 and 4.6

Table 4.5 Simulation results after apply T1MG + GT2SI in ORL

Test	Mean	Std	Max
1	0.9663	0	0.975
2	0.9390	0	0.975
3	0.9325	0	0.975
Mean	0.9459	0	0.975

Table 4.6 Simulation results after apply T2MG + GT2SI in ORL

Test	Mean	Std	Max
1	0.9459	0.012	0.975
2	0.9491	0.006	0.975
3	0.9502	0.012	0.975
Mean	0.9484	0.0098	0.975

respectively. The mean obtained after applying T1MG + GT2SI is 0.9459, while that using the T2MG + GT2SI is 0.9484.

4.2.2 Simulation Results Using Cropped Yale Database

The results achieved when the Cropped Yale database is used are shown in Tables 4.7 and 4.8. In Table 4.7 can be observed an example of the parameters obtained in each test.

In Tables 4.8 and 4.9 we can find the results of 3 simulations by using the T1MG and T2MG in Cropped Yale database respectively. In this case the use of T1MG + GT2SI produced a recognition rate of 0.9976 and with the T2MG + GT2SI a recognition rate of 0.9995 is obtained.

With the results obtained by the MNN using the ORL and Cropped Yale databases, it can be determined that the proposed method of GT2SI in combination with the pre-processing performed in the images with the edge detectors allows improving the recognition rates.

4.2.3 Comparison of Results with Other Aggregation Operators

An aggregation operator has the main function of combining information from different sources. In Table 2.2, the most common aggregation operators are presented, as well as the methods of integration of responses used to integrate the information coming from the different modules of the MNN.

Table 4.7 Simulation results after apply T1MG + GT2SI in 27 tests on the Cropped Yale database

Test	Mean	Std	Max
1	0.9811	0.0172	1
2	0.9784	0.0177	1
3	0.9784	0.0177	1
4	0.9811	0.0172	1
5	0.9811	0.0172	1
6	0.9784	0.0177	1
7	0.9811	0.0172	1
8	0.9811	0.0172	1
9	0.9811	0.0172	1
10	0.9811	0.0172	1
11	0.9784	0.0177	1
12	0.9784	0.0177	1
13	0.9811	0.0172	1
14	0.9811	0.0172	1
15	0.9784	0.0177	1
16	0.9811	0.0172	1
17	0.9811	0.0172	1
18	0.9811	0.0172	1
19	0.9811	0.0172	1
20	0.9784	0.0177	1
21	0.9784	0.0177	1
22	0.9811	0.0172	1
23	0.9811	0.0172	1
24	0.9784	0.0177	1
25	0.9811	0.0172	1
26	0.9811	0.0172	1
27	0.9811	0.0172	1
Mean	0.9802	0.0173	1

Table 4.8 Simulation results after apply T1MG + GT2SI in Cropped Yale

Test	Mean	Std	Max
1	0.9993	0.0039	1
2	0.9991	0.0020	1
3	0.9944	0.0142	1
Mean	0.9976	0.0067	1

Table 4.9 Simulation results after apply T2MG + GT2SI in Cropped Yale

Test	Mean	Std	Max
1	0.9987	0.0012	1
2	1	0	1
3	1	0	1
Mean	0.9995	0.0004	1

Table 4.10 Aggregation operators using T1MG in ORL

Method	Mean	Std	Max
SI	0.8773	0.0538	0.9625
IT2SI	0.8812	0.054	0.9625
GT2SI	0.9459	0	0.975
Choquet	0.9002	0.0501	0.9875
IT2 Choquet	0.9333	0	0.9375
The winner takes it all	0.885	0.0548	0.95
Arithmetic mean	0.915	0.0548	0.9875
Harmonic mean	0.875	0.0415	0.9125
Geometric mean	0.93	0.0401	0.975
Weighted arithmetic mean	0.9007	0.0477	0.9875
Owa	0.9016	0.0471	1

This section provides a comparison of the recognition rates achieved by the face recognition when different aggregation operators are applied. To make a comparative analysis of the behavior of the results, for this study case the operators were implemented: Sugeno integral, the interval type-2 Sugeno integral, GT2SI, the Choquet integral, the interval type-2 Choquet integral, the winner takes it all, arithmetic mean, harmonic mean, geometric mean, weighted arithmetic mean and OWA.

The results of applying the aggregation operators using T1MG in the ORL database are presented in Table 4.10. It is possible to appreciate that the proposed operator presents better results with a value of 0.9459 with respect at the rest of the aggregators and the value obtained for each aggregation method can be appreciated in Fig. 4.5.

In Fig. 4.6, it is possible to observe the behavior of the results obtained in Table 4.11 according to the values calculated with the aggregation operators using the ORL database with IT2MG.

The mean of using the aggregation operators with T1MG and IT2MG in the Cropped Yale database are presented in Tables 4.12 and 4.13 respectively. The use of GT2SI with T1MG produces a recognition rate of 0.9509 that is better than the rest of the operators used (Fig. 4.7).

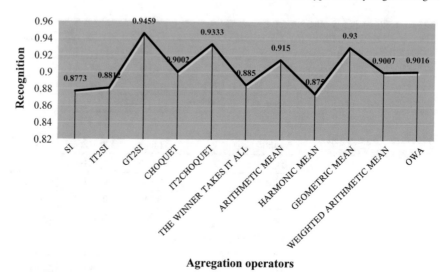

Fig. 4.5 Comparison for T1MG in the ORL database

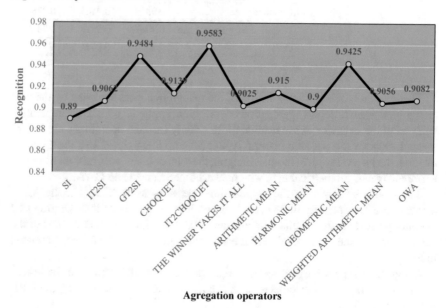

Fig. 4.6 Comparison T2MG in the ORL database

Table 4.11 Aggregation operators using T2MG in the ORL

Method	Mean	Std	Max
SI	0.89	0.0538	0.95
IT2SI	0.9062	0.054	0.9
GT2SI	0.9484	0.0098	0.975
Choquet	0.9139	0.0482	0.963
IT2 Choquet	0.9583	0	0.963
The winner takes it all	0.9025	0.0479	0.938
Arithmetic mean	0.915	0.0548	0.975
Harmonic mean	0.9	0.0405	0.938
Geometric mean	0.9425	0.0243	0.963
Weighted arithmetic mean	0.9056	0.0451	0.975
OWA	0.9082	0.0452	0.975

Table 4.12 Aggregation operators using T1MG in Cropped Yale

Aggregation operator	Mean	Std	Max
SI	0.8227	0.0538	1
IT2SI	0.8136	0.054	1
GT2SI	0.9976	0.0067	1
Choquet	0.9509	0.0543	1
IT2 Choquet	0.9837	0.0401	1
The winner takes it all	0.9263	0.0736	1
Arithmetic mean	0.9816	0.0288	1
Harmonic mean	0.8026	0.0483	0.8816
Geometric mean	0.9789	0.02	1
Weighted arithmetic mean	0.892	0.0553	1
Owa	0.8899	0.0591	1

Table 4.13 Aggregation operators using T2MG in ORL

Aggregation operator	Mean	Std	Max
SI	0.8364	0.0718	1
IT2SI	0.8386	0.0698	1
GT2SI	0.9995	0.0004	1
Choquet	0.9509	0.0543	1
IT2 Choquet	0.9974	0.0059	1
The winner takes it all	0.9263	0.0736	1
Arithmetic mean	0.9816	0.0288	1
Harmonic mean	0.8026	0.0483	0.8816
Geometric mean	0.9789	0.02	1
Weighted arithmetic mean	0.892	0.0553	1
Owa	0.8899	0.0591	1

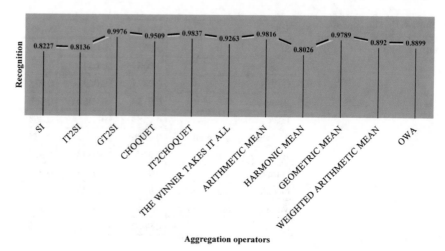

Fig. 4.7 Comparison T1MG in the Cropped Yale database

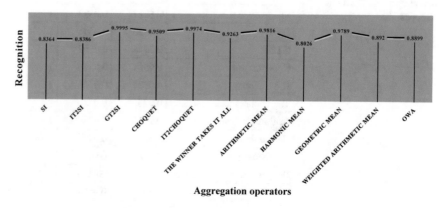

Fig. 4.8 Comparison T2MG in the Cropped Yale database

We can observe that in the Cropped Yale database, when applying the pre-processing of images through the IT2MG edge detector, better results are obtained with the proposed GT2SI aggregation operator than the other used operators.

The recognition rates obtained with each aggregator are presented in Table 4.13 and in Fig. 4.8 it is possible to visually appreciate such a behavior.

With the tests carried out and the results obtained, it can be seen that the proposed method can handle large amounts of uncertainty and considerably improve the results with respect to other operators.

References

1. Martínez GE, Melin P, Mendoza OD, Castillo O (2015) Face recognition with a Sobel edge detector and the Choquet integral as integration method in a modular neural networks. Design of intelligent systems based on fuzzy logic, neural networks and nature-inspired optimization, pp 59–70
2. Melin P, González CI, Castro JR, Mendoza O, Castillo O (2014) Edge-detection method for image processing based on generalized type-2 fuzzy logic. IEEE Trans Fuzzy Syst 22(6):1515–1525
3. González CI, Melin P, Castro JR, Castillo O, Mendoza O (2016) Optimization of interval type-2 fuzzy systems for image edge detection. Appl Soft Comput 47:631–643
4. González CI, Melin P, Castro JR, Mendoza O, Castillo O (2016) An improved sobel edge detection method based on generalized type-2 fuzzy logic. Soft Comput 20(2):773–784
5. Martínez GE, Mendoza O, Castro JR, Melin P, Castillo O (2013) Generalized type-2 fuzzy logic in response integration of modular neural networks. IFSA World Congress and NAFIPS annual meeting (IFSA/NAFIPS)
6. Martinez GE, Mendoza O, Melin P, Gaxiola F (2016) Comparison between Choquet and Sugeno integrals as aggregation operators for modular neural networks. FUZZ-IEEE, pp 2331–2336
7. Martínez GE, Mendoza DO, Castro JR, Melin P, Castillo O (2017) Choquet integral and interval type-2 fuzzy Choquet integral for edge detection. In: Nature-inspired design of hybrid intelligent systems, pp 79–97
8. Martinez GE, Mendoza O, Castro JR, Rodríguez Díaz A, Melin P, Castillo O (2016) Comparison between Choquet and Sugeno integrals as aggregation operators for pattern recognition. NAFIPS, pp 1–6
9. Martínez GE, Mendoza O, Castro JR, Rodríguez-Díaz A, Melin P, Castillo O (2015) Response integration in modular neural networks using Choquet Integral with Interval type 2 Sugeno measures. NAFIPS/WConSC, pp 1–6
10. Martínez GE, Mendoza O, Castro JR, Melin P, Castillo O (2014) Choquet integral with interval type 2 Sugeno measures as an integration method for modular neural networks. WCSC, pp 71–86
11. Melin P, Mancilla A, Lopez M, Mendoza O (2007) A hybrid modular neural network architecture with fuzzy Sugeno integration for time series forecasting. Appl Soft Comput 7(4):1217–1226
12. Ontiveros E, Melin P, Castillo O (2018) High order α-planes integration: a new approach to computational cost reduction of general type-2 fuzzy systems. Eng Appl AI 74:186–197
13. Melin P, Sánchez D, Castillo O (2012) Genetic optimization of modular neural networks with fuzzy response integration for human recognition. Inf Sci 197:1–19
14. Sánchez D, Melin P, Castillo O (2017) Optimization of modular granular neural networks using a firefly algorithm for human recognition. Eng Appl AI 64:172–186

Chapter 5
Conclusions and Future Work on the Generalized Type-2 Fuzzy Sugeno Integral

In this work, the main goal was to develop a method to extend the fuzzy Sugeno integral using generalized type-2 fuzzy logic. The proposed method was developed and it was demonstrated that it works properly in two very different applications. The proposed fuzzy Sugeno integral using generalized type-2 fuzzy logic was based on previous works, like in [1–8].

In the case of the (MG + GT2SI) presented in Sect. 4.1, according to the results obtained it was shown that when combining a traditional method of edge detection with the method of aggregation of the Sugeno integral extended with generalized type-2 fuzzy logic, the performance at the time of the edge identification was much better than with traditional methods [7, 9–11], so it can be determined that the use of GT2SI is viable to improve the performance image processing methods based on gradient measures.

On the other hand, the experimental results given by the integration of the MNN (MNN + GT2SI) which were presented in Sects. 4.2 and 4.3, we can conclude that although the SI, IT2SI and GT2SI aggregation methods are able to combine the information satisfactorily and produce the correct result, the combination of the SI with generalized type-2 fuzzy system is able to handle a higher level of uncertainty than IT2SI and SI, what you can see in the results shown in the Sect. 4.2.3 for the cases of faces recognition using the ORL and Cropped Yale databases.

With the development of this work it can be determined that when we have a system or problem in which there is a high level of uncertainty, it is possible to make use of generalized type-2 fuzzy systems, since, due to their nature, they may have greater uncertainty management, especially compared to type-1 fuzzy systems and interval type-2 fuzzy systems. The results obtained demonstrate that there are considerable advantages when using them, however, for the moment the proposed method of GT2SI cannot be applied in real-time systems due to the processing power that is required to be able to carry out that homework.

As future work, it is proposed to use optimization techniques in order to explore and find the appropriate parameters for the generalized type-2 membership function, so that the performance of the proposed system is optimal, as well as perform an

© The Author(s), under exclusive license to Springer Nature Switzerland AG 2020
P. Melin and G. E. Martinez, *Extension of the Fuzzy Sugeno Integral Based on Generalized Type-2 Fuzzy Logic*, SpringerBriefs in Computational Intelligence,
https://doi.org/10.1007/978-3-030-16416-4_5

exploration to determine if the number of alpha cuts used [12] in pattern recognition and edge detection applications was the one that allows to take the most advantage of the aggregation operator.

It is also intended to make use of the operator in other application problems where it is necessary to perform numerical aggregation, as well as in problems where there is an amount n of diverse sources of information. In addition, it is proposed to create a system by means of which it is possible to analyze the data contained in each source of information so that, based on them, the fuzzy densities assigned to each source can be automatically generated. It is also suggested to use the operator in other application problems where it is necessary to perform numerical aggregation, as well as in problems where there is an amount n of diverse sources of information [13, 14].

This document presents an approximation of the GT2SI, however it is still necessary to analyze if it is possible to carry out the calculations in some other way.

References

1. Martínez GE, Mendoza O, Castro JR, Melin P, Castillo O (2013) Generalized type-2 fuzzy logic in response integration of modular neural networks. In: IFSA World Congress and NAFIPS annual meeting (IFSA/NAFIPS)
2. Martinez GE, Mendoza O, Melin P, Gaxiola F (2016) Comparison between Choquet and Sugeno integrals as aggregation operators for modular neural networks. FUZZ-IEEE, 2331–2336
3. Martínez GE, Mendoza O, Castro JR, Melin P, Castillo O (2017) Choquet integral and interval type-2 fuzzy Choquet integral for edge detection. In: Nature-inspired design of hybrid intelligent systems, 79–97
4. Martinez GE, Mendoza O, Castro JR, Rodríguez Díaz A, Melin P, Castillo O (2016) Comparison between Choquet and Sugeno integrals as aggregation operators for pattern recognition. NAFIPS, 1–6
5. Martinez GE, Mendoza O, Castro JR, Rodriguez-Diaz A, Melin P (2015) Oscar castillo. In: Response integration in modular neural networks using Choquet integral with interval type 2 Sugeno measures. NAFIPS/WConSC, 1–6
6. Martínez GE, Melin P, Mendoza OD, Castillo O (2015) Face recognition with a Sobel edge detector and the Choquet integral as integration method in a modular neural networks. In: Design of intelligent systems based on fuzzy logic, neural networks and nature-inspired optimization, 59–70
7. Martínez GE, Mendoza O, Castro JR, Melin P, Castillo O (2014) Choquet integral with interval type 2 Sugeno measures as an integration method for modular neural networks. WCSC, 71–86
8. Melin P, Mancilla A, Lopez M, Mendoza O (2007) A hybrid modular neural network architecture with fuzzy Sugeno integration for time series forecasting. Appl Soft Comput 7(4):1217–1226
9. Melin P, González CI, Castro JR, Mendoza O, Castillo O (2014) Edge-detection method for image processing based on generalized type-2 fuzzy logic. IEEE Trans Fuzzy Syst 22(6):1515–1525
10. González CI, Melin P, Castro JR, Castillo O, Mendoza O (2016) Optimization of interval type-2 fuzzy systems for image edge detection. Appl Soft Comput 47:631–643
11. González CI, Melin P, Castro JR, Mendoza O, Castillo O (2016) An improved sobel edge detection method based on generalized type-2 fuzzy logic. Soft Comput 20(2):773–784
12. Ontiveros E, Melin P, Castillo O (2018) High order α-planes integration: a new approach to computational cost reduction of general Type-2 Fuzzy systems. Eng Appl of AI 74:186–197

13. Melin P, Sánchez D, Castillo O (2012) Genetic optimization of modular neural networks with fuzzy response integration for human recognition. Inf Sci 197:1–19
14. Sánchez D, Melin P, Castillo O (2017) Optimization of modular granular neural networks using a firefly algorithm for human recognition. Eng Appl AI 64:172–186

Appendix A

Pseudocode of the Sugeno Integral

INPUT: Number of information sources n; information sources x_1, x_2,..., x_n; fuzzy densities of information sources $M_1, M_2, \ldots, M_n \in (0, 1)$.

OUTPUT: Sugeno integral $h(\sigma(x_1), \sigma(x_2), \ldots, \sigma(x_n))$.

Step 1: Calculate λ finding the root of the equation

$$f(\lambda) = \left\{ \prod_{i=1}^{n} (1 + M_i(x_i)\lambda) \right\} - (1 + \lambda) = 0$$

Step 2: Fuzzify variable x_i.

$$D_i = \left\{ x, \mu_{Di}(x) | x \in X \right\}, \mu D_i(x) \in [0, 1]$$

Step 3: Reorder M_i with respect to $D(x_i)$ in descending order.
Step 4: Calculate fuzzy measures for each data with (2.19).
Step 5: Calculate Sugeno integral with (2.20).
Step 6: OUTPUT Sugeno integral.

Pseudocode of the Generalized Type-2 Sugeno Integral

Input: Number of information sources n, information sources $D(x_i)$, Fuzzy densities of the information sources $M(x_i) \in (0,1)$.

Output: Generalized type-2 Sugeno integral h.

© The Author(s), under exclusive license to Springer Nature Switzerland AG 2020
P. Melin and G. E. Martinez, *Extension of the Fuzzy Sugeno Integral Based on Generalized Type-2 Fuzzy Logic*, SpringerBriefs in Computational Intelligence,
https://doi.org/10.1007/978-3-030-16416-4

Step 1: Evaluate each $D(x_i)$ and $M(x_i)$ with the function

$$\tilde{\mu}(x, u) = trigausstype2(x, u, [a_1, b_1, c_1, a_2, b_2, c_2, \rho])$$

Step 2: Calculate the α_i cuts for each $M(x_i)$ and $D(x_i)$

$$\mu(M_{iL\alpha_i}(x_i)), \mu(M_{iR\alpha_i}(x_i))$$
$$\mu(D_{L\alpha_i}(x_i)), \mu(D_{R\alpha_i}(x_i))$$

Step 3: Calculate λ and the α_i cuts for λ_L and λ_R using

$$f(\lambda_{L\alpha_i}) = \left\{ \prod_{i=1}^{n}(1 + M_{iL\alpha_i}(x_i)\lambda_{L\alpha_i}) \right\} - (1 + \lambda_{L\alpha_i}) = 0$$

$$f(\lambda_{R\alpha_i}) = \left\{ \prod_{i=1}^{n}(1 + M_{iR\alpha_i}(x_i)\lambda_{R\alpha_i}) \right\} - (1 + \lambda_{R\alpha_i}) = 0$$

Step 4: Calculate the Fuzzy measures $\mu_{L\alpha_i}(A_i)$ and $\mu_{R\alpha_i}(A_i)$

$$\mu_{L\alpha_i}(A_1) = \mu_{L\alpha_i}(x_1)$$
$$\mu_{L\alpha_i}(A_i) = \mu_{L\alpha_i}(x_i) + \mu_{L\alpha_i}(A_{i-1}) + \lambda_{L\alpha_i}\mu_{L\alpha_i}(x_i)\mu_{L\alpha_i}(A_{i-1})$$
$$\mu_{R\alpha_i}(A_1) = \mu_{R\alpha_i}(x_1)$$
$$\mu_{R\alpha_i}(A_i) = \mu_{R\alpha_i}(x_i) + \mu_{R\alpha_i}(A_{i-1}) + \lambda_{R\alpha_i}\mu_{R\alpha_i}(x_i)\mu_{R\alpha_i}(A_{i-1})$$

Step 5: For each α_i calculate the Sugeno integral $h(\tilde{\sigma}_{\alpha_i})$
$$h(\tilde{\sigma}_{\alpha_1}, \tilde{\sigma}_{\alpha_2}, \ldots, \tilde{\sigma}_{\alpha_n}) = \sqcup_{l=1}^{n}(\sqcap[h_{L\alpha_1}, h_{R\alpha_1}], \sqcap[h_{L\alpha_2}, h_{R\alpha_2}], \ldots, \sqcap[h_{L\alpha_n}, h_{R\alpha_n}])$$
where

$$\tilde{\sigma}_{\alpha_i} = \sqcup_{i=1}^{n}\left(\sqcap\left(\left[\mu(D_{L\alpha_i}(x_i)), \mu_{L\alpha_i}(A_i)\right], \left[\mu(D_{R\alpha_i}(x_i)), \mu_{R\alpha_i}(A_i)\right]\right)\right)$$

Step 6: Calculate the GT2SI $h = \overset{sup}{\underset{i}{}} (h(\tilde{\sigma}_{\alpha_i}))$

Appendix B

A. Function to Define the Sugeno Integral

```
function integral=integralsugeno(fuentes,g,L)
    gf_ord=sortrows([g',fuentes'],2);
    medidas=medidas_sugeno(gf_ord(:,1),L);
    [dd dd2]=size(fuentes);
    x=zeros(dd2);
    x(1)=((gf_ord(1,2)-0)*gf_ord(1,1));
    sum=x(1);
    for i=2:dd2
        x(i)=((gf_ord(i,2)-x(i-1))*gf_ord(i,2));
        sum=sum+x(i);
    end
     integral=sum;
end
```

B. Function to Define the GT2SI Edge Detector

```
close all;
clear all;

%%%%%%%%%%%%%% Obtaining synthetic and reference images
f=(rgb2gray(imread('donut_color[1].png')));
fref2=double(imread('donut_wire_black.png'));
```

```
% Assignment of diffuse densities to each gradient
dX=[0.5 0.4 0.9 0.7];

% Lambda calculation
lambda=calc_lambda21(dX(1,:));

% obtaining the gradient with the function
[G1 G2 G3 G4]=absdeltamg(f);
ss=max(max([G1; G2; G3; G4]));
G1=G1/ss;
G2=G2/ss;
G3=G3/ss;
G4=G4/ss;

%Calculate the GT2SI
[i,j]=size(G1);
im=zeros(i,j);
im2=zeros(i,j);
for x=1:1:i
    for y=1:1:j
        %Generalized Sugeno integral
        integral3=ISDDG_MM(dX,[G1(x,y),G2(x,y),G3(x,y),G4(x,y)]);
        %Sugeno integral
        integral4=integralSugeno([G1(x,y),G2(x,y),G3(x,y), G4(x,y)],dX,
lambda);
        im4(x,y)=integral4;
    end
end

%GM
G=((G1+G2+G3+G4));
im=G;
```

C. Function to Define the Interval Type-2 Sugeno Integral for Three Modules of a MNN using MG Edge Detector in the Preprocessing

```
pn_archivo=load(['matriz_',base,'_',detector,'_',num2str
(cant_muestras),'.mat']);
pn_completo=pn_archivo.matriz_bordes;
```

```
[ren,col]=size(pn_completo);
objetos=1:k;
objetivo=repmat(objetos,1,N/folds);
size(objetivo)
FOUDD=0.6;
[xx,yy]=size(dX);
ML=[];
MR=[];
dX=[];
for x=0.1:0.4:0.9
    for y=0.1:0.4:0.9
        for z=0.1:0.4:0.9
                dX(aaa,1)=x;
                dX(aaa,2)=y;
                dX(aaa,3)=z;
                aaa=aaa+1;
        end
    end
end

lambdaL=[];
lambdaR=[];
for ii=3:29;
ML=[];
MR=[];
% calculation of the left interval of the diffuse densities
   for x=1:yy
        if dX(ii-2,x)>FOUDD/2
            ML(x)=dX(ii-2,x)-FOUDD/2;
        else
            ML(x)=0.001;
        end
   end
 % calculation of the right interval of the diffuse densities
    for x=1:yy
        if dX(ii-2,x)<(1-FOUDD/2)
            MR(x)=dX(ii-2,x)+FOUDD/2;
        else
            MR(x)=0.9999;
        end
    end
 % ML
 % MR
 % calculate lambda (only Sugeno and Choquet)
```

```
        lambdaL(ii-2)=calc_lambda21(ML);
        lambdaR(ii-2)=calc_lambda21(MR);
for i=1:folds
        archivo=['fold',num2str(i),'_cross',base,'_',detector,'.mat'];
        load(archivo);
        pn_test=pn_completo(:,logical(coltest));
         X1=[];    X2=[];    X3=[];
        Tic
        %%%TRAINING
        %%%test with morphological gradient 111*91 =10101
        mod1=simula_imagenes(net1,pn_test(1:3367,:),k,N/folds); %Module1
        mod2=simula_imagenes(net2,pn_test(3368:6734,:),k,N/folds);%
Module2
        mod3=simula_imagenes(net3,pn_test(6735:10101,:),k,N/folds);%
Module3
                X1=[X1 mod1];
                X2=[X2 mod2];
                X3=[X3 mod3];
% Normalize the responses of the MNN with a triangular membership %function
                sX11=trimf(X1,[min(min(X1)),max(max(X1)),max(max(X1))]);
                sX22=trimf(X2,[min(min(X2)),max(max(X2)),max(max(X2))]);
                sX33=trimf(X3,[min(min(X3)),max(max(X3)),max(max(X3))]);
                % Determine the diffuse densities for each information source
                [ren,col]=size(sX1);
                % determine the interval for each information source
                sX1L=[];%
                sX2L=[];%
                sX3L=[];%
                sX1R=[];%
                sX2R=[];%
                sX3R=[];%
                FOUFI=0.2;
                  for r=1:ren
                      for c=1:col
                           if(sX1(r,c)>FOUFI/2)
                             sX1L(r,c)=sX1(r,c)-FOUFI/2;
                           else
                             sX1L(r,c)=0;
                           end
                           if(sX1(r,c)<(1-FOUFI/2))
                             sX1R(r,c)=sX1(r,c)+FOUFI/2;
                           else
                             sX1R(r,c)=1;
                           end
                           if(sX2(r,c)>FOUFI/2)
```

```
                        sX2L(r,c)=sX2(r,c)-FOUFI/2;
                    else
                        sX2L(r,c)=0;
                    end
                    if(sX2(r,c)<(1-FOUFI/2))
                        sX2R(r,c)=sX2(r,c)+FOUFI/2;
                    else
                        sX2R(r,c)=1;
                    end
                    if(sX3(r,c)>FOUFI/2)
                        sX3L(r,c)=sX3(r,c)-FOUFI/2;
                    else
                        sX3L(r,c)=0;
                    end
                    if(sX3(r,c)<(1-FOUFI/2))
                        sX3R(r,c)=sX3(r,c)+FOUFI/2;
                    else
                        sX3R(r,c)=1;
                    end
                end
            end
            IXL=[];
            IXR=[];
            % Calculation of the Sugeno integral by intervals
            for r=1:ren
                for c=1:col
                    %data with intervals
                    XL=[sX1L(r,c) sX2L(r,c) sX3L(r,c)];
                    XR=[sX1R(r,c) sX2R(r,c) sX3R(r,c)];
                    integralL=integralSuegnoT2(XL,ML,lambdaL(ii-2));
                    integralR=integralsuenoT2(XR,MR,lambdaR(ii-2));
                    %%% Sugeno integral
                    IXL(r,c)=integralL;
                    IXR(r,c)=integralR;
                end
            end
            IX=[];
            % The final decision will be the individual with the greatest
%value of the IS
            IX=(IXL+IXR)/2;
```

Index

© The Author(s), under exclusive license to Springer Nature Switzerland AG 2020
P. Melin and G. E. Martinez, *Extension of the Fuzzy Sugeno Integral Based on Generalized Type-2 Fuzzy Logic*, SpringerBriefs in Computational Intelligence,
https://doi.org/10.1007/978-3-030-16416-4

Printed in the United States
By Bookmasters